Galápagos

Galápagos

John Kricher

A Natural History

Princeton University Press

Princeton and Oxford

Requests for permission to reproduce material from this work should be
sent to Permissions, Princeton University Press.

First published by Smithsonian Institution Press in 2002

First Princeton edition, 2006

Library of Congress Control Number 2005935314
ISBN-13: 978-0-691-12633-3
ISBN-10: 0-691-12633-X

British Library Cataloging-in-Publication Data is available

This book has been composed in Clearface

Printed on acid-free paper. ∞

pup.princeton.edu

Printed in the United States of America

10   9   8   7   6   5   4   3

To Martha Steele
"T. O. M."

Contents

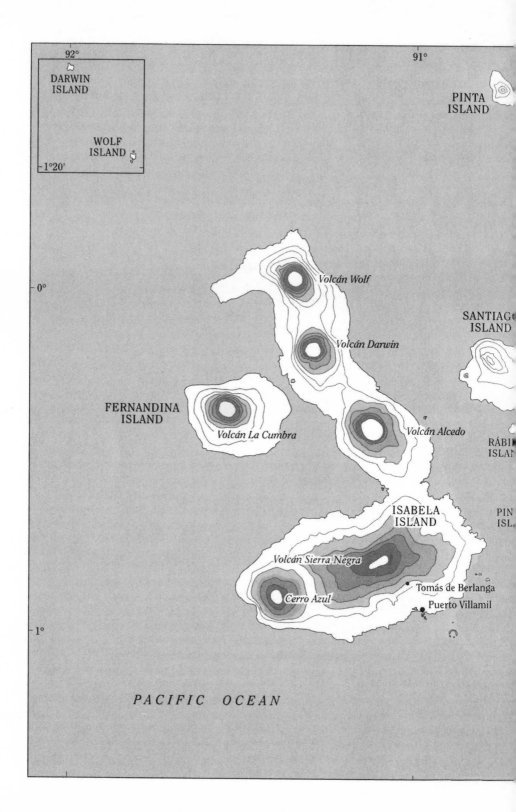

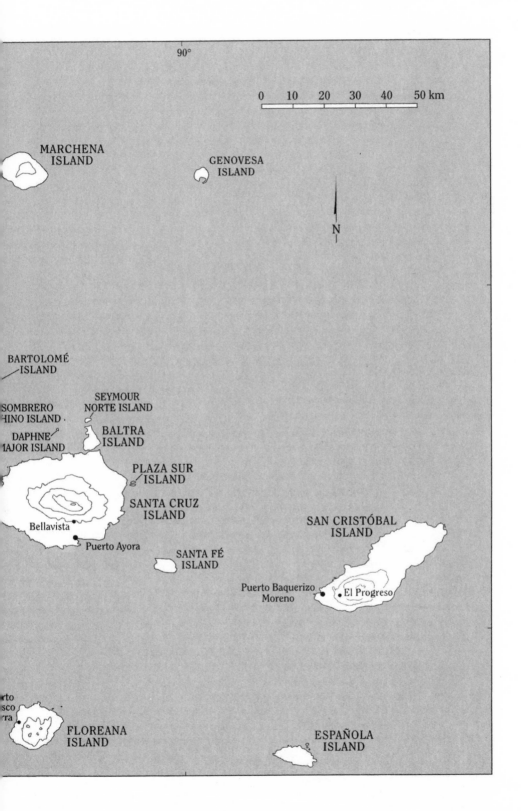

Preface

This is a book about two topics, a place and a process, both of which will forever be associated in history. The place is unlike any other, a unique group of islands, once so shrouded with mystery that they were referred to as the enchanted isles but are now simply called the Galápagos. The process is biology's major paradigm, organic evolution, the intellectual key that unlocks an understanding of how life is and how it got to be that way. These topics, place and process, are melded in the modern study of natural history and ecology.

It was once all so simple. It was believed that animals looked the way they looked because they were created to look that way. Same with plants. Organs such as eyes, to take the classic example, were so complex that they were analogous to watches, so perfectly structured that the only reasonable explanation was that they were designed in a purposeful way. After all, you cannot have a watch without a watchmaker, an analogy used metaphorically by William Paley in his influential book *Natural Theology* (1802), a book celebrating Christianity through natural history. Before becoming convinced of evolution, Darwin found Paley's book gave him "as much delight as did Euclid." He subsequently changed his mind.

When Darwin was a youth, it was obvious to him that fish were created to swim and that birds were created to fly. Giraffes were made with long necks to reach up and grab tasty leaves on tall trees. Humans whose faith and culture resided in Judeo-Christian teachings readily accepted creationism because it seemed an obvious, logical way of explaining nature and, above all, was philosophically satisfying and even comforting. Beyond creationism's explanatory powers, based as they were on nothing more than simple observations and belief, creationism gave nature purpose. Everything had to have been created for a reason. Nothing was left to chance.

For some, creationism is still philosophically satisfying, so much so that these individuals choose to ignore or even ridicule the single most important discovery ever made in biology: Organisms really do evolve.

Big ideas require lots of time to ferment, and evolution is a very big idea. By the nineteenth century many thinkers had entertained the notion that life was not always in its present form, that it had a past, an actual history. Advances in physics and astronomy had by then shown beyond reasonable doubt that Earth was part of a system of planets that each revolved in elliptical orbit around the sun—a solar system. No longer did Earth occupy center stage in the cosmos. Geology was rapidly converging on the notion that Earth was very old, much older than had been assumed. Now, instead of the age of the planet being thousands of years, it was believed to be in the millions of years, and probably the high millions at that. And the rocks of Earth spoke not only of great antiquity but also of change, of seas rising and retreating, mountains uplifting and eroding. Biologists were finding compelling similarities among extant organisms. The bones of the human arm could be seen, though in different proportions, in the wing of a bird or the foreleg of a dog. Embryos as different as those of fish and fowl appeared to be essentially identical early in their development. And then there were the fossils, petrified bones, outlines of life-forms preserved in rocks, organisms now extinct. Fossils showed clear anatomical similarities with modern, extant organisms.

Aristotle had conceived of a *Scala Naturae,* a great chain of being, but he envisaged it to be a scale of increasing perfection among the living creatures of the world, from the most perfect gods to the least perfect soil and slime. Aristotle's was a scale of creation, a static, unchanging scale. Now the rapidly emerging discipline of paleontology, the study of ancient life, was suggesting a profoundly different kind of chain, a great chain that runs not through levels of perfection but, instead, through time, linking the present inhabitants of the planet with species long gone. What was needed was someone to take this disparate assemblage of information and synthesize it. What did it all mean?

It meant, of course, evolution, and the person most responsible for figuring it out was Charles Darwin. Of course there were others. Darwin's grandfather Erasmus wrote of evolution in a lighthearted book of poems and essays, *Zoonomia.* In the year Darwin was born, 1809, Jean Baptiste de Lamarck published *Philosophie Zoologique,* a treatise on evolution. But Darwin went far, far beyond what anyone had previously articulated. It was Darwin who convincingly argued that all organisms that lived were linked by heredity and time with all that had ever lived—related in a magnificent, complexly branched metaphorical bush, a common descent of life. And it was Darwin who articulated and popularized a mechanism for how life-forms could change in response to changes in their environments, a mechanistic process, natural selection, blind to purpose but nonetheless producing complexity and order.

How did Darwin manage to do what he did? In large part, he succeeded because he had been to a very special place. The place was the catalyst. Here he collected some fascinating animals and plants. It was this place, the Galápagos Islands, and those animals and plants that were ultimately responsible for convincing Darwin that life was not created as is, but evolved. Had he never been there, would he have conceived of natural selection? Would he have postulated descent with modification? Would he have come to believe in transmutation of species and thus become an evolutionist?

The Galápagos Islands get your attention. Where else in the world can you go and look at flamingos marching in their collective mating display and then, to cool off from the intense equatorial sun, go swimming with penguins? Where else do colorful fishes of coral reefs mingle with big-eyed fur seals? Where else do large, austere lizards forage in the sea and small, nondescript birds rewrite all of biology?

As you read this book, I introduce you to the Galápagos Islands and to evolutionary biology. I begin with my first impressions of the islands, impressions gathered with other ecotourists on a trip typical of most that tour the islands. I then recount the odd human history of the Galápagos, islands that, for some of their history, attracted less than sterling examples of humanity. The third chapter is an overview, a basic primer on the Galápagos: the climate, geology, ecology, and natural history. I describe some of the factors that influence which kinds of organisms eventually populate remote volcanic islands such as the Galápagos. The fourth chapter is about Charles Darwin's life as it was affected by his visit to the Galápagos. This is followed by a discussion of the fascinating geology of the islands—how they were made, how they will age, and how they will die. Chapter 6 introduces the giant tortoises, whose saddlelike shells gave the islands their common name, Galápagos. Chapter 7 is devoted to the other reptilian tenants of the islands, most particularly the unique marine and land iguanas. The next two chapters describe the Galápagos avifauna, beginning with the seabirds and followed by the land birds. The latter includes the famous finches, now named for Charles Darwin, that are the "smoking gun" of evolutionary biology. Chapter 10 deals with many kinds of animals: crabs, sea lions, whales, sea turtles, and still more birds, the inhabitants of the shoreline and seas around the islands. The chapter titles are taken from Darwin's comments scattered throughout his widely read account of his voyage around the world, *The Voyage of the* Beagle. The penultimate chapter, Chapter 11, is my assessment of conservation efforts that are ongoing in attempting to preserve the Galápagos and some of the problems that these efforts face. The final chapter is an island-by-island look and handy guide for anyone planning to visit the Galápagos. This chapter tells you what you will find on the various islands and where to find it.

The Galápagos Islands found their way into my psyche long ago when my parents, who, luckily for me, always encouraged my burgeoning interest in natural history, presented me with an oversized book called *The Wonders of Life on Earth* on my birthday. This volume, published in 1960 when I was sixteen years old, was devoted entirely to a series of wonderfully written, profusely illustrated chapters about Charles Darwin's life and about how he came to write *On the Origin of Species*. The cover bore an evocative painting of a dome-shelled Galápagos giant tortoise against a lush verdant landscape with a brilliant male vermilion flycatcher on a branch in the foreground. Chapter 1, "A Laboratory of Evolution," was all about the Galápagos Islands and featured numerous striking photographs of the islands and their inhabitants. But what stood out above all were two stunning three-page foldouts, an artist's renderings of the island wildlife. One foldout depicted Darwin's finches—little black and brown birds of varying sizes and beak shapes—one picking at cactus pulp, one downing fruits, several cracking seeds. An imposing yellow land iguana serenely dominated the center of the foldout.

But it was the other foldout that somehow found its way into my very essence. It was of Punta Espinosa on Fernandina, then identified by the aristocratic-sounding name of Narborough. The painting was dominated by a marine iguana colony and depicted dozens of the incredible creatures in all their collective reptilian glory, with

Technicolor red Sally Lightfoot crabs scampering atop some of them. Frigatebirds were in the air, sea lions on the beach, and a huge flightless cormorant was feeding its young, dominating the right side of the painting. Austere in the background loomed the immense rim of Volcán La Cumbre, the giant shield volcano.

I had to go there. I just had to see *that place*, be at that spot, above all others on Earth. And I've been. And it was better than the painting. I can only hope that this book does these magnificent islands, and the unique life-forms they harbor, the justice that they deserve.

Acknowledgments

This book was originally published in hardcover by Smithsonian Institution Press, and I am grateful to the late Peter Cannell for his encouragement and support. I also thank Vincent Burke for adding his considerable editorial skills to the manuscript. I am most grateful to Robert Kirk of Princeton University Press for his enthusiasm in producing a paperback edition of the book.

Peter Alden, who has led more than two dozen tour groups to the Galápagos, worked with the Massachusetts Audubon Society in basically inventing what is now called *ecotourism*. He has pioneered routes around the world that are now well-known and frequented by individuals and tour companies alike. I doubt that his knowledge of global biogeography and natural history is matched by anyone. In spite of a plethora of opportunities to visit places of far greater biodiversity, Peter keeps returning to the Galápagos. I understand why. He has my thanks for his willingness to read the book when in draft form and, most of all, for making it possible for me to visit the Galápagos for the first time.

I am deeply grateful to Dr. Edward H. Burtt, Jr., who meticulously read the entire manuscript and made detailed comments throughout. At one point he suggested a reference I should consult. He apologized for not providing a full citation but noted that he was reading the chapter while awaiting a flight at Miami International Airport. He cited the gate number. Jed's considerable editorial skills as well as his encompassing knowledge of the Galápagos resulted in significant improvement.

I also thank Dr. Jerome A. Jackson and Martha Steele, both of whom read the entire manuscript and whose suggestions made it more readable throughout. My colleague Dr. Betsey Dyer lent a critical eye to the geology chapter, and I thank her for making it stronger.

My close friend William E. "Ted" Davis, Jr., who Peter Alden describes as my "sidekick," accompanied me on my first trip to the Galápagos and read a draft of the entire manuscript, and I thank him for his suggestions.

Drs. John Faaborg, Patricia Parker, and Rob Fleischer were very helpful in sharing information about the genetics of the Galápagos hawk, and I am grateful to each.

I thank Marcia Grimes, reference librarian at Wheaton College, for her help in researching various scientific papers.

Some of the photographs that enhance this book were taken by my good friend Bruce Hallett. Bruce, whose photography skills are matched only by his birding prowess and quick wit, was kind enough to send me a selection of his photographs, and I thank him for allowing me the liberal use of them.

This book really became a kind of team project when I traveled to the Galápagos in June of 1997 with a group of eleven close personal friends. Martha Steele, who has a deep fondness for the Galápagos, took it upon herself to handle the complex planning process that is a prerequisite to any successful trip. In addition to the trip being enlightening and helping focus my plan for how I would author a book about the Galápagos, it was immense fun. Fond recollections of daily excursions on the islands are matched by memories of evening conversations on board the *San Jacinto,* the Southern Cross overhead, the Big Dipper "upside down" to the north, and Comet Hale-Bopp near the horizon. The only way to improve on a trip to a place such as the Galápagos is to take along a bunch of your best friends.

My wife, Martha Vaughan, who is my best friend, and who was one of the eleven members of the "Darwin Group," helped research and copyedit the manuscript throughout its production. Her constant support and encouragement made my task easier, and this book is much improved from her efforts. Thanks, Martha. Like the rest of our life together, nice collaboration.

1

Nothing Could Be Less Inviting

Charles Darwin did not like what he saw. It was September 17, 1835, and the HMS *Beagle* had carried Darwin, then twenty-six years old, from his secure home in rural England to a forbidding archipelago of remote, arid black volcanic islands in the equatorial Pacific Ocean, nearly 600 miles (about 1,000 km) west of the South American country of Ecuador. The journey had already been long and arduous, especially the repeated, depressing bouts of seasickness. It had begun almost four years earlier, on a cold December 27, 1831, and by the time he set foot on the Galápagos, Charles had seen almost enough natural history to last him a lifetime. Almost. Indeed, after finally returning from the *Beagle* journey on October 2, 1836, Charles would never again leave his native England. He had, by then, gained sufficient experience to inspire thoughts that would culminate almost a quarter of a century later in a book that would change all biology, to say nothing of Western philosophy. Darwin's epiphany derived in no small measure from what he witnessed while on these remote and desolate volcanic islands. But he would not know that until he had returned to England. And on that hot day in mid-September of 1835, very far from home, Darwin was not favorably impressed. "Nothing could be less inviting than the first appearance," he wrote of Chatham Island (now called San Cristóbal), one of the larger islands of an archipelago named for the lumbering giant tortoises, the galápagos, that then abounded on the various islands.

My experience was different from Darwin's. I first saw San Cristóbal on May 23, 1987, some 152 years after Darwin. To my eyes, thanks in large part to Darwin's influence, nothing could have looked more inviting. I was about to visit the Rosetta stone of evolutionary biology. The MV *Santa Cruz*, considerably larger and more comfortable than the HMS *Beagle*, had just dropped anchor at this, the most eastern of the Galápagos Islands. The island itself did, indeed, appear rugged. Steep precipices of black lava rock abruptly met the shoreline, framed in a backdrop of eroded volcanic cones. A lingering rainy season plus a cool drizzle, which I soon learned was called *garúa*, had provided enough moisture that the vegetation that covered the upper regions of the volcanic slopes remained in leaf, the soft greens oddly reminiscent of the Scottish highlands, not bleak, barren, and seemingly leaf-less as when Darwin visited in the height of the dry season. The highest peaks were immersed in low-lying clouds. There was a smooth, sandy beach extending for con-siderable distance along the shoreline and a dense forest of short leafless trees just beyond it.

Along with about a dozen other ecotourist passengers, I boarded a Zodiac, called a *panga* when used in the Galápagos, for my introduction to San Cristóbal. *Panga,* a word used extensively throughout the Galápagos, describes any small flat-bottomed boat commonly used in the islands, whether a Zodiac or a dory. Our initial objective was to circumnavigate Cerro Brujo, an old tuff cone with deep fissures and small caves carved by the unceasing energy of an ever-restless sea.

As we approached the ragged ebony rocks, it was low tide, and the narrow inter-tidal zone was well exposed. Crabs were abundant on the rock faces, immature and adult Sally Lightfoots, the strikingly scarlet adults adding a splash of bright color where there otherwise would be none. A small committee of stoic marine iguanas, the gargoyles of the Galápagos, gazed at us, unmoving. The cliffside was inhabited by some blue-footed boobies, large marine birds related to northern gannets and streamlined for plunge diving into the sea. With such brilliant blue webbed feet, it was easy to see how they got their common name. They were roosting on the nar-rowest of the ledges, a common perching site as revealed by accumulated "white-wash," the droppings from the birds. Slender frigatebirds, known for their piratical habit of pursuing other birds (especially boobies) and forcing them to drop their food, were suspended on stiff wings in the air overhead, ever vigilant for an oppor-

MV *Santa Cruz* in Tagus Cove, Isabela

tunity to highjack a few more calories. The largest birds were brown pelicans, groups of them flying serenely, almost at water level, vaguely resembling pteranodons, long extinct pterosaurs from the Cretaceous period, when dinosaurs were still in vogue.

Indeed, the whole scene was distinctly antediluvian, and the combination of the tall escarpment of blackened rocks, tenanted by primitive-looking birds, overly active crabs, and robust iguanas, reminded me of Skull Island, the mythical land where dinosaurs still lived and a very large ape named King Kong once ruled. Yet I knew that Skull Island was actually inspired by the Indonesian island of Komodo, where huge Komodo dragon lizards are found. No giant apes in either place.

The panga approached the shoreline closely, waves crashing around us, the channel just sufficiently wide to allow us passage, the volcanic rocks of Cerro Brujo looming ominously. Acorn barnacles the size of golf balls populated the waterline, and dozens of crabs scurried about just beyond the barnacles. One drab, grayish gull stood sentry duty alone on a rock at the water's edge. It was a lava gull, one of the many Galápagos endemic species, its evolutionary history probably linked to that of the much more widely distributed laughing gull. *Endemism,* which refers to a species that has evolved locally and is normally not found elsewhere, is a common occurrence throughout the archipelago. There certainly was a resemblance between the lava gull and the laughing gull. The lava gull looked like a laughing gull that had gotten trapped in a chimney and emerged covered with soot. A single wandering tattler, a well-named shorebird whose global perambulations take it to many out-of-the-way islands, was teetering characteristically as it methodically searched among the algae for arthropods to help it refuel.

Having seen Cerro Brujo, it was time to set foot on the Galápagos. The panga took us a safe distance away from the rocky area to where the beach was smooth and inviting, and we made a wet landing (it's as it sounds, you get your feet wet) along a stretch of sandy beach. As I was preparing to exit the panga, a small, sparrowlike brown bird, with short rounded wings and a tail so stubby as to resemble that of a fledgling, flew hurriedly past me. It certainly wasn't exotic in appearance, far from it. I could sympathize with Darwin, who paid little attention to these nondescript feathered creatures when he first encountered them. But this was no ordinary sparrow. In fact it wasn't a sparrow at all, but a *Geospiza fuliginosa,* otherwise known as a small ground finch, one of a group of fourteen species of birds that now bear Darwin's name.

Darwin's finches have become one of the linchpins in the theory of evolution, a process that Darwin called "descent with modification." From a single common ancestor that long ago colonized the Galápagos, some thirteen species have evolved, some with large, powerful, seed-crushing bills, others with thinner bills, one with a forcepslike bill so much like that of a wren that the bird was once mistaken for one, and by Darwin himself. There is a fourteenth species of the Darwin's finch on Cocos Island, off Costa Rica. As Darwin eventually wrote, "Seeing this gradation and diversity of structure in one small, intimately related group of birds, one might really fancy that from an original paucity of birds in this archipelago, one species had been taken and modified for different ends" (from *The Voyage of the* Beagle). Shortly after returning from the *Beagle* voyage, Darwin, in large part because of the finches, began the construction of an argument about how life got to be what it is, an argument that would occupy his deepest thoughts for the remainder of his years.

With the benefit of Darwin's hindsight, I searched through a flock of small ground finches to see if any other species were present. Soon I saw two ground finches *(G. fortis)* that were noticeably larger in body size, with proportionally huskier bills. They didn't look that much different from the small ground finches. Still, they were different, importantly different. Among the dozen or so finches that were actively flitting among the palo santo trees along the shoreline, some had black bills; some had light brown bills. Black bills signify that the birds are breeding. Darwin's finches continue to be outstanding examples of the evolutionary paradigm of natural selection, the survival of the fittest.

Two other birds drew my attention away from the finch flock. Both were Chatham mockingbirds—greenish eyed, brownish on the back, with blackish faces—which, like the finches, are species endemic to the Galápagos. Before Darwin paid much attention to the finches, he noticed that the mockingbirds, or *mocking-thrushes* as he called them, differed a bit from island to island. Indeed, the Chatham mockingbird occurs nowhere else on the planet but San Cristóbal. There are three other mockingbird species on the Galápagos, one of which, the Hood mockingbird, is restricted to Española (formerly Hood), another of which, the Charles mockingbird, is restricted to tiny Champion Island off Floreana (formerly Charles), where it was once common. Darwin first thought these birds to be mere varieties, but when he learned they were separate species, he found it curious that the Creator would mold such slightly different creatures to occupy such similar islands. Another explanation suggested itself, the one Darwin eventually choose to believe. Once he came to that belief, that the mockingbirds and the finches evolved, he never looked back.

Chatham mockingbird, San Cristóbal

Just ahead of me where the sandy beach rose to meet the vegetation, three lava lizards were scurrying about in the dry leaf litter. As with other Galápagos creatures, these reptiles were essentially tame, allowing very close approach before moving. Seven species of lava lizards live on the archipelago, another case of evolutionary divergence. One of the three individuals before me, a male, was distinctly larger and more colorful than the other two. Each of the smaller lizards had a bright orange throat, a characteristic of females. The male was doing energetic push-ups, its reptilian brain directing it to "impress the females and maybe you'll get lucky." It was certainly trying.

As I walked between the densely growing but still leafless palo santo trees, I began to notice a distinct odor reminding me somewhat of turpentine. I asked one of our naturalist guides about it, and he said I was smelling the holy tree. I learned later that *palo santo* translates to "holy stick" and that the species itself *(Bursera graveolens)* is genetically similar to myrrh and frankincense. Exudate from the tree is burned in churches as incense. Unlike the immense plant diversity of the South American mainland, the Galápagos flora is anything but vast, and palo santo is one of the most abundant plant species, dominating the dry zone vegetation that occurs at low elevations throughout the islands. Among the leafless palo santo forest I saw a different kind of tree, one with bright green shiny leaves more typical of the wet tropical forests with which I was intimately familiar. The guide cautioned me not to touch it because it can cause skin eruptions similar to those from poison ivy. It was the manzanillo *(Hippomane mancinella),* or poison apple tree, its resin so toxic that no part of it is really safe to touch. Yet another unusual plant attracted my attention: the tall, columnar candelabra cactus, a species in the genus *Jasminocereus* that reminded me of the organ pipe cacti that grow in the Sonoran Desert in extreme southern Arizona and Mexico. What an odd collection of vegetation. But then again, oddness is anything but odd on these islands.

Back on the beach, I was looking at some flowering beach morning glory *(Ipomoea pes-caprae)* when I noticed peculiar markings in the sand that led from the upper beach to the water's edge. They were tracks made by a green sea turtle as she labored up the beach to lay her eggs and then returned to the sea. Nearby a young male California sea lion was reclining on the strand. Like most other Galápagos creatures, it had no reaction to either my close approach or my rapidly clicking camera. With utter nonchalance, it ambled casually into the sea, where it swam off in the direction of Kicker Rock, an eroded volcanic memento that sits just off the west coast of San Cristóbal. As I walked back to the panga, it was late in the afternoon, and blue-footed boobies were diving headfirst just offshore. Blue-foots feed closer to shore than either of the other two booby species that nest on the Galápagos, and it was clearly their dinnertime.

It was nearing our dinnertime as well, and my group had reboarded the panga and was about to depart San Cristóbal for the *Santa Cruz* when our naturalist guide noticed that one of the passengers was holding a feather. It looked to be a flight feather from a gull, or perhaps a booby. He asked where she had gotten the feather, and she told him it was just lying on the beach. She thought it would make a suitable memento of San Cristóbal. The guide, polite but firm, reminded her, as well as the rest of us, that conservation rules are strict on the Galápagos, all of which are protected as a national park of Ecuador. Nothing from any plant or animal or rock or mineral is to be removed. The passenger, obviously embarrassed but also some-what irritated, asked what harm it would do to take one feather. The guide gave the obvious answer: The islands host thousands of tourists annually, and without strict enforcement of conservation policies, there would be an immense cumulative human impact on the islands, just from tourists alone. The passenger, in a bit of a huff, tossed the feather back on the beach. The guide, patient and polite but persis-tent, asked where, exactly, she had first picked up the feather. She pointed down the beach about 20 yards. The guide then asked her to please retrieve the feather and take it back to where she had found it. Those were the rules. She complied. We waited until our now contrite companion was back aboard, and then the panga sped us over the waves to the *Santa Cruz* in time for a lavish dinner.

I couldn't help but think how astonished Darwin would be if he could have wit-nessed this afternoon of ebullient ecotourism in a place where "Nothing could be less inviting than the first appearance."

Selected References

Darwin, C. 1845. *Journal of Researches into the Natural History and Geology of the Countries Visited by H.M.S.* Beagle *round the World.* [Later published as *The Voyage of the* Beagle. 1906. London: J. M. Dent & Sons.]

2

First by *Bucaniers,* and Latterly by Whalers

A s with almost any isolated archipelago, it can never be known for certain just who was the first person to set foot on the islands and how and when that event occurred. The Galápagos are archeologically depauperate, except for some scattered pottery of uncertain origin that, though ancient, may have been transported there in relatively recent times. There are no archeological ruins to explore or human remains to analyze. Thor Heyerdahl, following a visit to the islands in 1953, argued that pre-Inca peoples could and did navigate to and from the Galápagos and used the islands for fishing. There is also a claim that an Incan ruler, Tupac Yupanqui, made landfall on the Galápagos sometime in the 1400s. There is no firm confirmation for either assertion.

The islands were isolated from human recorded history until February 1535, when there is an unambiguous record of their discovery by the Bishop of Panama, Fray Tomás de Berlanga, whose ship was blown off course as he made his way from Panama to Peru. In other words, the bishop may well have found the Galápagos in exactly the same manner as the ancestors of the marine and land iguanas that had colonized many centuries earlier and whose descendants the bishop did not fail to notice on his brief, unintentional visit. Like Darwin, the bishop was not favorably impressed by the islands, finding them rocky and desolate, though he did notice the unusual tameness of the fauna and he commented about the large size of the tortoises. Unfortunately, the bishop made landfall on the islands during the dry season when they have little or no fresh water. Several of his party succumbed to dehydration before a meager amount of spring water was located. However, the majority did eventually make it to Peru. And the Galápagos made it onto the map.

The archipelago was included in a world atlas published in 1570, prepared by a Flemish cartographer, Abraham Ortelius. Ortelius gave the archipelago the name Insulae de los de Galopegos, the name inspired by the saddle shape of the shells of some of the giant tortoises. This name has endured even though the islands were officially named Archipiélago del Ecuador after their annexation by Ecuador on February 12, 1832, and renamed Archipiélago de Colón in 1892 to honor the four hundredth anniversary of the discovery of the Americas by Columbus (whose Spanish name is Cristóbal Colón and who never sailed the Pacific Ocean). Official Ecuadorian maps notwithstanding, and with all due respect to Christopher Columbus, this archipelago will forever be the Galápagos.

Following their discovery, the islands became a sort of refuge for seafarers, many, probably most, of dubious distinction. During the seventeenth century scores of buccaneers and privateers used the islands as a convenient base from which to attack Spanish ships, the gold-carrying galleons that transported the spoils from the Incan conquest. The pirates, a rough crowd indeed, also occasionally attacked the coastal towns scattered along Ecuador and Peru. A Spanish pirate, Deigo de Rivadeneira, is credited with giving the islands their first name, though it was not used by Ortelius, preparer of the atlas cited above. Deigo attempted to land but couldn't because of the unusually strong and confusing interisland currents. So powerful were the waters that swirled around the islands that Deigo even entertained the notion that the islands themselves were somehow adrift in the maelstrom. Attributing his apparent lack of seamanship to divine causation, Deigo named the islands Las Islas Encantadas, which translates to "the bewitched isles."

Deigo notwithstanding, most other pirates managed to land on the islands with little difficulty. The original Robinson Crusoe, a Scotsman whose actual name was Alexander Selkirk, famous for having spent four years on the island of Juan Fernandez before being rescued, retreated to the Galápagos after a raid on the Ecuadorian city of Guayaquil. It seems Crusoe's (Selkirk's) rescuer, Captain Woodes Rogers, was a pirate of some repute, and Selkirk was honored to be his second mate.

Piracy eventually ceased, but the Galápagos were still extensively used by mariners throughout the eighteenth and nineteenth centuries. The quest was no longer for gold but for living commodities, especially whales and sea lions. In 1793, the British captain James Colnett made detailed maps of the archipelago for use by Great Britain's whaling fleet. Sperm whales were plentiful in the sea near the Galápagos, and both California sea lions and Galápagos fur seals (which are really sea lions) abounded on the islands. The quest for these animals brought many vessels and resulted in the demise of many a pinniped.

Because the islands figured so prominently as a port for sailing ships, some lonely but resourceful mariners erected a barrel on the island of Floreana that served as an eclectic sort of post office. Floreana, whose official name is Santa Maria, was originally named Charles but is commonly called Floreana, an example of the confusion that sometimes occurs with the names of the individual islands. Ships bound for England or eastern North America would put in at Post Office Bay, as the spot soon became known, to pick up mail from the barrel, while those headed elsewhere would post mail, hoping it would be transported back home and hand delivered by a well-meaning sailor. This barrel has been repeatedly replaced over the years, and the tradition continues (now among ecotourists) even today. But, alas, it is not very reliable. None of my mail made it to the intended destinations.

In 1813, the United States Navy frigate *Essex* arrived at the Galápagos and made observations of volcanic eruptions on Fernandina and Isabela. The *Essex* was in hostile waters. The War of 1812 was in full swing, which was an excellent excuse for the *Essex* to pretty much annihilate the British whaling fleet. But the ship's real legacy on the islands has little to do with the war. Captain David Porter released four goats on Santiago (then called James Island and also known as San Salvador) so that they might graze. The hardy and enterprising ungulates escaped, and they and their descendants have outcompeted tortoises and other native animals ever since. When I first visited the Galápagos in 1987, I was told that Ecuadorian marines were brought to some of the islands to eradicate the goats by shooting them. The goats, which succeed far better than humans in negotiating the thick Galápagos underbrush, won, and the marines went back to Ecuador. Goats, as well as pigs, dogs, cats, and donkeys, remain a problem, though many of these unwelcome animals have been successfully eradicated from several of the islands.

In August 1813, two officers from the *Essex* had a serious altercation aboard ship and decided to settle their dispute in a duel, to be held on James Island. That they did, and a midshipman named Cowan was killed by a marine lieutenant named Gamble. The ship's company buried Cowan on the island with full military honors. Gamble likely saluted his corpse. Oddly, his grave has not been located since. There are many such tragic human stories that color the early history of the enchanted islands.

The United States had relatively little to do with the Galápagos until the Second World War. At that time the United States believed the Galápagos to be a strategic area for the protection of the Panama Canal and received permission to build an airstrip on the island of South Seymour, now called Baltra. This airstrip, the construction of which wreaked havoc on the ecology of the island (the soldiers shot land iguanas for amusement), is still used for flights to and from the islands. Today a second airstrip located on San Cristóbal is also used heavily for tourist flights. The United States left the islands in 1946, after the war ended, but not before attempting to lease or buy the archipelago, an attempt that failed.

Throughout their history, the Galápagos Islands offered more than merely a safe haven for sailors and their ships. The islands, at least the larger ones, had water (particularly during the rainy season), wood, and plenty of food in the form of birds, iguanas, sea lions, and fur seals, as well as the almost ubiquitous giant tortoises. The extraordinarily tame sea lions and fur seals that populate most of the islands provided an easy source of meat and hide. The innocent, uncomprehending creatures would watch restlessly but idly while one after another of their companions was bashed to death by sailors. But it was the tortoises that were most valuable to seafarers. Because of their low metabolic rate, these immense reptiles could be kept alive on a ship with no food or water for months at a time. Clearly thousands, and some accounts suggest well over one hundred thousand, were taken by whalers and sealers, stacked on ships, and consumed for their meat and oil. William Dampier, who sailed with the pirate captain Woodes Rogers, mentioned above, described the tortoises as, "especially large and fat; and so sweet, that no Pullet eats more pleasantly." He also considered the Galápagos doves and iguanas to be quite tasty.

No one really attempted to live on the Galápagos until about 1807 (the precise date is uncertain), when a man named Patrick Watkins became a self-proclaimed permanent resident of Floreana. He was apparently not a model citizen: He allegedly traded potatoes and tobacco for rum as well as other forms of alcohol that he used to maintain himself in a relatively constant state of advanced inebriation. In appearance he was extraordinarily unkempt, his tattered clothes clinging to his body like so many limp rags. In personality he was decidedly unpleasant. He had the annoying habit of periodically capturing visiting sailors at gunpoint and making slaves out of them. For reasons known only to him, he apparently left the islands with some of his prisoners in a stolen boat in 1809 and eventually reached Guayaquil, allegedly after killing, and some accounts suggest eating, all of the other men whom he took with him. He never returned to the islands but was instead eventually incarcerated. Probably just as well.

Floreana and the rest of the islands remained free of human inhabitants until 1832, three years before Darwin's visit. At this time a small colony was formed on Floreana, founded by an Ecuadorian general named Jose Villamil. Today a small village, not on Floreana but on the southern tip of Isabela, is his namesake. The Floreana colony met with modest initial success and totaled somewhere between 250 and 300 when Darwin visited in 1835. Like many such colonies, however, this one was composed largely of societal misanthropes, especially convicts. Probably because of a combination of harsh treatment and overall difficult conditions, a revolt occurred, and the colony failed. Though the people did not stay, their domestic animals remained, causing much ecological damage. The tortoises that once roamed abundantly on the island were reduced to a rarity, a story unfortunately repeated on numerous other islands as humans brought goats, cattle, pigs, dogs, cats, rats, and donkeys.

Other colonies appeared. On San Cristóbal (then Chatham), General Villamil founded a small colony after abandoning his charges on Floreana. This colony also failed. In 1869, another colony, mostly composed of slaves, was begun on San Cristóbal, organized by a man named Manuel Cobos. Though Cobos named his colony Progreso, his idea of progress seemed to involve liberal use of flogging as well as execution by dismemberment, perhaps the reason why his colonists eventually ganged up on him and slaughtered him with machetes, an apparent retaliation for his having done the same thing to several of their company. Though Cobos's colony failed to endure, its name did, and today there is the town of El Progreso on San Cristóbal.

Other attempts at colonization, mostly using slave labor, proved less than successful, and thus it wasn't until the twentieth century that lasting, successful colonies were finally established. Today there are modest villages on Santa Cruz, San Cristóbal, Isabela, and Floreana, supporting a total population of approximately 16,000. The human population in the islands is increasing rapidly. A census from 1950 indicated a total of 1,346 inhabitants. In a half century, that population has increased nearly twelve times. The most visited settlement is clustered around Academy Bay on Santa Cruz, also the location of the Charles Darwin Research Station (founded in 1964) and the Galápagos National Park Service. The next substantial settlement is on San Cristóbal, at Puerto Baquerizo Moreno and El Progreso. It is on this island that children of the Galápagos attend the Carlos Darwin School, complete with an austere bust of the famous man overlooking the playground. On the southern end of Isabela, there is a small fishing settlement at Puerto Villamil. Finally, there is a small human population living on Floreana, most of whom farm the highlands, but one resident was famous, a true celebrity in Galápagos modern history. The closest thing to a town on Floreana is Black Beach, where there are scattered houses and a school. It is here that you find what was once a small guesthouse.

Until she died on March 21, 2000, at the age of ninety-five, Margret Wittmer resided on Black Beach and, for many years, ran the guesthouse. Frau Wittmer lived on Floreana for sixty-eight years, having emigrated from her native Germany along with her late husband Heinz when they were newlyweds in 1932. Inspired in part by

William Beebe's colorful account of the islands, the Wittmers decided to make a go at pioneering, bringing little with them from Germany. Living initially in a volcanic cave, the former real estate of pirates, the Wittmers persevered against very daunting odds, teaching themselves the diverse skills necessary for survival and eventually establishing a comfortable home and a successful farm, raising a small family, and entertaining numerous visitors, including many international dignitaries. And, last but not least, the Wittmers were a central part of a strange Galápagos story that makes most soap operas look tame in comparison.

When the Wittmers colonized Floreana, they had but one neighbor, also a German. His name was Dr. Friedrich Ritter, and he moved to Floreana with his second wife, Dore, in 1929. Dr. Ritter fancied himself quite the philosopher and intended to rewrite all of Western philosophy, an ambitious project. Being trained in medicine, he also had a pragmatic side and was worried over the lack of dental care on the remote, totally isolated island on which he and his wife were to be the only humans. He thus took the precaution of having all of his teeth extracted prior to leaving Germany. He believed in good exercise, lots of outdoor activity, nudity (but who was to see him anyway?), and vegetarianism. Until he ate the bad chicken. But I'm getting ahead of myself.

The Wittmers and Dr. Ritter did not get on well initially. Ritter was put off by the invading Wittmers, all two of them, and perhaps thought the island was becoming much too crowded, given that the population had suddenly doubled when the Wittmers arrived. But an uneasy tolerance was established, to the point where Dr. Ritter somewhat reluctantly assisted in the difficult birth of Frau Wittmer's first child.

Then along came the baroness, one Eloise Bosquet de Wagner Wehrborn of Vienna, who intended to set up a resort and also took the liberty of proclaiming herself empress of Floreana, entitled to all the rights and privileges that accompany royal status. The baroness, who was in the habit of carrying a sidearm and a whip, liked giving orders more than making friends. She was accompanied by two gentlemen, Lorenz and Philippson, both apparently brought along to be her lovers. Poor Lorenz fell out of favor with the rambunctious baroness and was soon reduced to a lowly and ill-treated servant, while Philippson won the honor of being her lover, presumably getting the better of the deal.

None of these inhabitants really got along very well. It was, to say the least, a tiny community of eccentrics, if indeed it can even be called a community. But it didn't last very long. Dr. Ritter abruptly died from eating spoiled chicken. Why he devoured the bird is unknown: He was a vegetarian; plus the chicken emitted a distinctly unpleasant odor. Ritter took care to boil it before eating it, insisting that the foul-smelling fowl would be rendered palatable by temporary immersion in scalding water. But it wasn't and he died. So much for Ritter.

Lorenz, terrified of the baroness, sought sanctuary with the Wittmers, who took him in, but he soon perished while exploring when he became shipwrecked on the island of Marchena, then known as Bindloe. His dehydrated corpse, mummified by the intense equatorial sun, was eventually found. So much for Lorenz.

The baroness and Philippson set sail for Tahiti, where she apparently thought she would start a resort and, of course, become the island's empress. They were never seen again, anywhere on the planet. So much for them.

These events occurred in such close sequence that the Wittmers were briefly under some suspicion of foul play, though not a shred of evidence to that effect was ever uncovered. Frau Wittmer has written a fascinating account of her life on Floreana, including, of course, her interactions with Ritter and the baroness, which has recently been reprinted and which I enthusiastically recommend.

The next phase of Galápagos history centers on the preservation of the natural history of the islands. Conservation laws were enacted to protect the islands in 1934, and the entire archipelago was declared an Ecuadorian national park in 1936. But it was not until 1958, at a meeting of the International Zoological Congress, that real conservation work began. A committee was established, chaired by Sir Julian Huxley, that was instrumental in establishing the Charles Darwin Research Station at Academy Bay on Santa Cruz. The Darwin Foundation was created in 1959, which also was the centennial year for the publication of Darwin's *On the Origin of Species*. To honor Darwin and the book, a penitentiary on Isabela was closed and destroyed. The station itself opened in 1962 and has since served as home base for visiting scientists as well as for training Galápagos natural history guides. The Charles Darwin Research Station is the nexus of conservation planning and action throughout the islands.

Modern residents of the archipelago, mostly Ecuadorian, make their livings from ecotourism, simple farming, fishing, and shop keeping. Approximately three out of four residents are considered "urban," while the rest are "rural." The Ecuadorian government is under pressure to permit more human immigration to the islands, but to do so would be a direct threat to the islands as a national park.

Notable Visitors to the Islands

Few would disagree that Charles Robert Darwin, then of Shrewsbury, England, will endure as the most notable visitor to the Galápagos. Much more is related about him in the pages that follow, but there is one thing worth mentioning here. In keeping with the colorful and often violent history of the islands, Darwin noted in his journal that while on Santiago (James), he found the skull of a captain of a sealing vessel who had been murdered some years earlier by his crew.

The first expedition that could be characterized as scientific in nature occurred in 1790, when a Sicilian named Alessandro Malaspina was sent to explore the islands by King Carlos IV of Spain. Little else in the way of scientific exploration took place until Darwin and the *Beagle* arrived forty-five years later, though in 1825 David Douglas, an eminent botanist, meticulously collected the plants that populate the islands. He noted the unique nature of the flora but never suggested any kind of evolutionary theory to account for the islands' natural history.

Louis Agassiz, leading naturalist of the United States, founder of the Museum of Comparative Zoology at Harvard University, world authority on ichthyology, and ardent opponent of Darwin's theories regarding evolution, visited the Galápagos for nine days in June of 1872, almost a half century after Darwin. For those who naively believe that a visit to the Galápagos Archipelago will automatically convert them to a belief in evolution, Douglas and Agassiz proved otherwise. In fairness, however, Agassiz visited the Galápagos only one year before his death at the age of sixty-six. Unlike Darwin, who was young and vigorous, and whose mind was still highly malleable when he explored the islands, Agassiz was frail, and his beliefs were more than a little firmly entrenched. He had very little to say in print concerning his impressions of the islands, though he did suggest in one weakly argued letter to a friend that his views concerning the truth of creationism were not shaken by seeing the Galápagos flora and fauna.

Other scientific expeditions were made to the islands after Darwin, but the most comprehensive was conducted by Rollo Beck, yet another opponent of Darwinism. Beck led a team of scientists from the California Academy of Sciences. This yearlong expedition, in 1905–1906, vaulted the California Academy into the position of world leader in Galápagos data and scientific publications, a position it essentially still maintains. The expedition was on-site in the Galápagos when the great earthquake of 1906 leveled much of San Francisco. The data collected on this expedition were so complete that even though the academy's museum was destroyed in the earthquake, its reputation and importance remained intact due entirely to the addition of voluminous Galápagos information once the expedition returned. The museum itself was soon rebuilt to house a complete collection of Galápagos specimens.

William Beebe, whose popular narratives *Galápagos—World's End* and *The Arcturus Adventure* are still widely read and enjoyed, made two trips to the islands, in 1923 and 1925. Beebe was a highly skilled and seasoned naturalist with a writer's talent for description and detail. Beebe's accounts of the various creatures he encountered, like Darwin's before him, are alive with enthusiasm for his subjects. In describing two juvenile marine iguanas tussling, Beebe wrote,

> One of them reared high upon his hind legs and tail, with forearms bent and claws outspread, and balanced for a moment before dropping upon his playmate. I was crouching almost flat so that the players were silhouetted against the sky, and they might have been any size in the world. The pose I have described was exactly that of the attacking Tyrannosaurus of Knight's splendid painting, clothing the great skeleton in the American Museum of Natural History, and again, as so often in these islands, I had the conviction of ancient days come once more to earth.

Beebe was referring to Charles R. Knight, who was the foremost artist of prehistoric animals. His well-known and oft-reproduced paintings and murals of dinosaurs and other extinct creatures are among the finest natural history art.

The noted British ornithologist David Lack was the first scientist to conduct detailed field studies on Darwin's finches. Indeed, Lack's study of the birds is largely responsible for promoting the collective name *Darwin's finches*. Lack's work, conducted from December 1938 to April 1939, was eventually published by the California Academy of Sciences in 1945, followed by a dramatic revision published in 1947. Like so many others who had gone before him, David Lack's initial description of the islands was anything but complimentary:

> The numerous recent travel books describing the Galápagos—the "Enchanted Isles"—had not sufficiently prepared us for the inglorious panorama. Behind a dilapidated pier and ramshackle huts stretched miles of dreary, greyish brown thornbush, in most parts dense, but sparser where there had been a more recent lava flow, and the ground still resembled a slag heap.

In spite of his initial less than enthusiastic reaction, Lack and his research team persevered, and his subsequent analyses of finch ecology and evolution helped establish a paradigm for ecological and evolutionary research, much of which has endured to the present day. Lack's efforts have been continued by others, most notably Peter and Rosemary Grant of Princeton University. The Grants and their many students have completed over two decades of highly detailed research on Darwin's finches.

Two highly distinguished writers from different centuries, Herman Melville and Kurt Vonnegut, have been inspired by their visits to the Galápagos.

Herman Melville, who is known best for his novel about an albino whale with a bad attitude and a sea captain with a grudge, came to the Galápagos in 1841, when he was in his twenty-first year. Melville had sailed over a year earlier out of New Bedford, Massachusetts, as a sailor aboard the *Acushnet,* the whaling vessel that provided the model for the fictional *Pequod*. His impressions of the islands, based largely on the tales he heard of the many eccentric outcasts that had resided therein, as well as his own impressions of the odd animals, resulted in a brief work about the islands titled "The Encantadas," published in 1856 in *Monthly Magazine*. Melville's view of the Galápagos seemed to center on the bizarre nature of both the nonhuman and human inhabitants. His description of the Galápagos penguin typifies his opinion of the islands as a land of nature's as well as society's rejects.

> [T]heir bodies are grotesquely misshapen, their bills short; their feet seemingly legless; while the members at their sides are neither fin, wing, nor arm. And truly neither fish, flesh, nor fowl is the penguin. . . .On land it stumps; afloat it sculls; in the air it flops. As if ashamed of her failure Nature keeps this ungainly child hidden away at the ends of the earth.

It has been suggested that Melville found the Galápagos intriguing because the islands may have represented a mocking imagery of nature, an imagery that he may have entertained of himself as a person. So if the Galápagos provided Darwin with eventual insight into evolution, perhaps the islands provided Melville with insight into his own persona. At any rate, the islands served both men well.

Much more recently Kurt Vonnegut published *Galápagos*, an entertaining futuristic novel about human evolution that is set in the Galápagos and that takes Darwin's theory almost too literally. Vonnegut basically posed the question that was presumably solved by some iguanas of the distant past—namely, when a terrestrial animal suddenly finds itself permanently isolated on the Galápagos, how does such a being eventually evolve into an aquatic creature? For the answer, at least for Vonnegut's answer, his book is recommended.

Selected References

Beebe, W. [1924] 1988. *Galápagos: World's End*. Reprint, New York: Dover Publications.
Berry, R. J. 1984. Darwin was astonished. *Biological Journal of the Linnean Society* 21:1–4.
Gottlieb, L. D. 1975. The uses of place: Darwin and Melville on the Galápagos. *Bioscience* 25:172–75.
Gould, S. J. 1983. Agassiz in the Galápagos. In *Hen's Teeth and Horse's Toes*. New York: Norton.
Jackson, M. H. 1985. *Galápagos: A Natural History Guide*. Calgary, Alberta, Canada: University of Calgary Press.

Lack, D. [1947] 1983. *Darwin's Finches*. Reprint, Cambridge, UK: Cambridge University Press.

Perry, R. 1972. *The Galápagos Islands*. New York: Dodd, Mead.

Slevin, J. R. 1959. The Galápagos Islands: A history of their exploration. *Occasional Papers of the California Academy of Sciences* 25:1–150.

Thornton, I. 1971. *Darwin's Islands: A Natural History of the Galápagos Islands*. Garden City, N.Y.: Natural History Press.

Vonnegut, K. 1985. *Galápagos.* New York: Delacorte Press.

Wittmer, M. 1996. *Floreana*. Quito, Ecuador: Ediciones Libri Mundi, 1989. Reprint, Trowbridge, Wiltshire, UK: Redwood Books.

3

This Archipelago

The Galápagos is an archipelago of recent volcanic origin located essentially on the Equator from 600 to 700 miles (approximately 1,000 km) from the coast of Ecuador and about 1,000 miles (1,600 km) from Panama. West of the Galápagos is nothing but the open Pacific Ocean for about 3,000 miles (4,800 km). The nearest other island is Cocos Island, a densely forested island about 600 miles (960 km) to the northeast of the Galápagos. The youngest major island, Fernandina (formerly Narborough), is less than a million years old. In total, the archipelago numbers somewhat in excess of seventy landmasses, though over forty of these are but tiny islets, some yet to be named, and most are rarely visited by people. In its entirety, the Galápagos archipelago covers about 27,968 square miles (45,000 km^2) of sea and comprises about 5,000 square miles (8,000 km^2) of landmasses. There are nineteen large islands; thirteen of these have an area in excess of 6 square miles (10 or more km^2). A typical tour of the Galápagos will include at least nine and some include up to fourteen of these islands. By far the largest island is Isabela (formerly Albemarle), whose area of 1,743 square miles (4,588 km^2) is nearly five times that of Santa Cruz (formerly Indefatigable), the next largest island (375 square miles [986 km^2]). Only seven islands exceed 38 square miles (100 km^2), and only nineteen exceed $\frac{1}{3}$ of a square mile (about 1 km^2).

The islands have a confusing nomenclature. Most were originally given English names. Some islands were named after prominent British citizens, such as the Duke of Albemarle, King Charles II, Admiral Viscount Duncan, R.N., and Admiral Sir John Narborough, and others, like Bindloe and Ewres, were named after pirates. Still others, like Beagle, Daphne, and Indefatigable, were named for ships. These names were essentially all dropped in favor of Spanish names when Ecuador annexed the Galápagos (see the following). However, all older accounts of the islands use the English names, so it is really helpful to know both. To make matters even more confusing, local custom sometimes uses the old English name even today. For instance, the island of Española is still usually called Hood, and Tower Island, whose official name is Genovesa, is almost always called Tower. Tour guides sometimes mix usage:

An English name may be favored for a particular island, and a Spanish name for another. A few islands are commonly called names other than either their English or Spanish names! In Santiago's case, its official name is San Salvador, it was originally named James, but everybody calls it Santiago.

All told, sixty-one of the Galápagos Islands have official names: thirteen major islands, six minor islands, and forty-two islets.

Unique Natural History

The Galápagos *are* natural history. No one goes there for the native people, the ruins, or the architecture because there is essentially none. Visitors go for the spectacle, the grandeur, and particularly the animals, many of which are extraordinarily tame. Though he traveled to many islands, Darwin felt the need to comment in his journal on the particular tameness of the Galápagos fauna. But there are also remarkable plants, a diverse array of marine life, and splendidly desolate lava fields on islands punctuated by tuff cones, calderas, and vistas that truly inspire. It is little wonder that visitors arrive by the thousands annually.

These once remote islands have in some ways become the signature islands for modern ecology and conservation. The words *ecotourism* and *Galápagos* are anything but strange bedfellows. It is rather ironic, actually, how an arid archipelago that failed to appeal to the likes of Melville and Darwin has since become a drawing card for ecotourists, a development that would in all likelihood be inconceivable to Darwin. Beginning in 1973, when regular flights from Ecuador to Baltra were put into service, masses of tourists have come annually to visit the Galápagos to enjoy the unique beauty of the islands as well as their remarkable natural history. Today somewhere in excess of seventy thousand tourists annually visit the archipelago, and there is constant pressure to increase that number.

The Galápagos Islands are home to remarkable creatures: giant tortoises, marine and land iguanas, colonies of seabirds and sea lions, flightless cormorants, the most northern species of penguin, and a notable group of finches. Almost all of these animals are uncommonly tame. You can walk past a tree on which sits a Galápagos hawk. The hawk, barely at arm's length, watches curiously but never moves. Penguins pay little attention to you as you swim by them. Boobies continue their mating dances oblivious of the many cameras clicking and camcorders whirring within a few feet of them. There are numerous landings that must be accomplished by walking over and around the sprawling sea lions that can't be bothered to get out of the way. Sit on the beach on Hood Island and curious mockingbirds will be pecking at your toes and trying to open your daypack. Marine iguanas by the dozen ignore you as you try to decide which group to photograph. Indeed, it's easy to become a good wildlife photographer on the Galápagos. You will rarely encounter such willing subjects in nature. Bring film, lots of film.

The Galápagos are like nowhere else. Situated directly on the Equator, the latitude at which tropical rainforests are typical on continents, the islands nonetheless have a generally variable climate, with long periods of intense aridity alternating with periods when deluges are not uncommon. And the islands provide a rugged, austere look, a remoteness that is further enhanced by the pervasive evidence of continual volcanic activity. As Darwin commented in his diary, "The country compared

Names of the Largest Galapágos Islands

Official Name	Original English Name	Commonly Used Name
Bainbridge	Bainbridge	Bainbridge
Baltra	South Seymour	Baltra
Bartolomé	Bartholomew	Bartolomé
Caldwell	Caldwell	Caldwell
Campeon	Champion	Champion
Coamano	Jensen	Jensen
Crowley	Crowley	Crowley
Darwin	Culpepper	Culpepper
Eden	Eden	Eden
Enderby	Enderby	Enderby
Española	Hood	Española
Fernandina	Narborough	Fernandina
Genovesa	Tower	Tower
Isabela	Albemarle	Isabela
Jardinero	Gardner-by-Charles	Gardner-by-Charles
Jardinero	Gardner-by-Hood	Gardner-by-Hood
Los Hermanos	Crossman	Los Hermanos
Marchena	Bindloe	Marchena
Mosquera	Daphne Major	Daphne Major
Onslow	Onslow	Onslow
Pinta	Abingdon	Pinta
Pinzón	Duncan	Duncan
Plaza Sur	South Plaza	Plaza Sur
Rábida	Jervis	Jervis
San Cristóbal	Chatham	San Cristóbal
San Salvador	James	Santiago
Santa Cruz	Indefatigable	Santa Cruz
Santa Fé	Barrington	Barrington
Santa Maria	Charles	Floreana
Seymour Norte	North Seymour	North Seymour
Sombrero Chino	None	Sombrero Chino
Tortuga	Brattle	Tortuga
Watson	Watson	Watson
Wolf	Wenman	Wenman

to what we might imagine the cultivated parts of the Infernal regions to be." Formed as a byproduct of one of the most geologically active areas on the planet, these youthful volcanic islands, created by the evacuation of material from deep ocean vents, and never connected with any continent, have been gradually colonized by plants and animals accidentally transported by air and water currents. These organisms, those that survived, formed the breeding stock from which most of the current species evolved. In general, the biodiversity of the Galápagos is very low compared with that of mainland areas, but this fact of nature, which was of no small interest to Darwin, can be easily understood. The islands have depended entirely on colonization and subsequent evolution to attain their present biodiversity.

Colonization

Islands such as the Galápagos offer both severe challenges and significant opportunities for would-be colonizers. A newly formed island harbors no competitors, predators, or parasites. On the other hand, there are very few sources of food or shelter in such a barren terrain. But life is tenacious. As the fictional character Ian Malcom said in Michael Crichton's bestseller *Jurassic Park,* "Life will find a way." But the fact remains that isolated islands are exactly that, isolated, and organisms cannot live there unless they get there. On the other hand, dispersal is a common adaptation in nature. Seeds, parts of plants, whole plants, or animals carried by wind, waves, or in some cases migratory birds are like lottery tickets of DNA. They move, often randomly, their fates controlled by the vagaries of atmospheric or oceanic currents. Most never make landfall, but a lucky few do, and some of those grow and survive and eventually thrive and reproduce. Plants find rootholds to secure enough minerals and moisture to persist and grow in the daily equatorial light. Sometimes the sea may carry whole plants that are broken loose from riverbanks when the rainy season floods the continental rivers. Small clumps of plants, miniature islands, float downriver and are carried into the open ocean. These clumps may also contain animals, both invertebrates and vertebrates, little natural arks floating out to sea. This process, called *rafting,* is a byproduct of the action of climate on mainland ecosystems, but it can result in major colonization events when given sufficient time.

One study of Galápagos plants by Duncan M. Porter in 1984 concluded that there had been 306 colonizations in about 3 million years, amounting to one successful colonization every one thousand years. This might seem to be a very low rate, but that is the essence of how colonization works. Given 3 million years, it is quite possible to clothe an oceanic island in green. Some animals such as small insects and spiders may be borne aloft in air currents and balloon their way to a fortuitous landfall on newly formed volcanic islands. Still other animals may be transported as eggs in the mud that cakes the feet of migrating shorebirds. Darwin, who is often mistakenly believed to have merely "thought" about evolution, performed meticulous experiments collecting mud from ponds, mud that could easily adhere to the feet of migrating birds, and seeing what sorts of plants grew out of it. Darwin's work on plant dispersal by migrating birds is described in *On the Origin of Species*.

Colonization is modest at its onset, but biodiversity does, eventually, increase. Colonization probabilities are anything but random. First consider geography. Organisms transported by water will move with currents, so current direction is all-important in determining which colonists might or might not arrive on the islands. The nearest mainland is also critical in supplying the islands with species. Distant mainlands are less likely to be sources. In the case of the Galápagos, Central and South America are likely sources, a mere 600–700 miles away. Ocean currents generally are favorable to transport from the Americas to the islands. Thus it is hardly surprising that of a total of 541 plant species 27 percent are from the lowland tropics of South America and 56 percent are from the Andean region. Others come from such places as the Caribbean, Mexico and Central America, and North America.

Biology also matters in colonization. Some seeds and seedlings, like those of mangroves, tolerate salt water with remarkable tenacity and can drift for long periods without harm, eventually finding an island on which to forge a roothold. Other plants perish quickly if conditions deteriorate. Reptiles can withstand long periods without food or water; their scale-covered skins resist desiccation. But amphibians, whose skins must remain moist, are very different kinds of animals. It is little wonder that reptiles are first-rate colonizers and amphibians are anything but. Mammals, like amphibians, are generally poor colonizers, but for a different reason. In the case of mammals it is metabolism that limits their dispersal powers. Mammals have high energy demands and must eat and drink far more frequently than reptiles. It is little wonder that of the four thousand or so mammals of the world only two sea lion species, two bat species, and two rice rat species have found their way to the Galápagos. The sea lions swam, one from the north, one from the south. The bats flew. The rats probably rafted from South America. Rats are tough, and they sometimes get lucky.

Endemism

Many Galápagos plants and animals occur nowhere else on Earth. Biogeographers and evolutionary biologists use the word *endemic* to describe a species that is restricted to a particular geographic area and nowhere else. The implication, of course, is that the species evolved there, though this need not be the case. A species may, in theory, colonize an area and then become extinct from the remainder of its former range. If such a history were not understood, the species would appear to be endemic. A case in point is the California condor *(Gymnogyps californianus)*, which was once widespread in North America but in recent times has been totally restricted to southern California. It is not California endemic, though it would appear to be.

The Galápagos have very high endemism. You'll not find *Amblyrhynchus cristatus,* the marine iguana, anywhere else. The same is true with *Creagrus furcatus,* the swallow-tailed gull, *Zenaida galapagoensis,* the Galápagos dove, and *Myiarchus magnirostris,* the Galápagos flycatcher, as well as numerous others. The plant species *Scalesia pedunculata, Lycopersicon cheesemanii,* and *Lantana peduncularis* are among the approximately 250 endemic plant species found on the archipelago, a total that represents about 45 percent of all plant species on the islands. One group of plants from the sunflower family in the genus *Scalesia* has diversified into fourteen species, five of which have distinct varieties on various islands, and this group is the vegetational equivalent to Darwin's finches. Endemism among Galápagos plants is unusual in that most endemism occurs among coastal and arid (lowland) zone plants, with much-reduced endemism among highland plants. But this odd distribution may be explained by climatic history (see the following).

There are twenty-two endemic land bird species of the twenty-nine species that regularly occur, an endemism percentage of 76 percent. As for seabirds, 26 percent of the species are endemic, including two endemic genera. A subspecies is a genetically recognizable population. The term *race* means essentially the same thing as *subspecies*. In 1984, David Snow and Brian Nelson found that, at the subspecies level, there is an endemism of 58 percent, making the seabird assemblage here more highly endemic than on any other collection of islands in the world. The frequently sighted little lava lizards of the genus *Tropidurus* have seven endemic species spread around the archipelago. Endemism is high for insects. In a recent study of Galápagos beetles, of which there are about four hundred species, it is estimated that the endemism rate is 67 percent.

Taken as a whole, the natural history of the Galápagos powerfully supports Darwin's insightful views about evolution. These islands are, in a real sense, the Eden of evolutionary biology. The odd assemblage of flora and fauna, the unique characteristics of the various species, the isolation of these life forms from mainland areas all provide intellectual grist for the evolution mill. Would Darwin have realized the reality of evolution had he not visited the Galápagos? Possibly. His colleague Alfred Russel Wallace did, and he had not visited the Galápagos, but he had spent eight years in Indonesia, then the Malay Archipelago, and had observed many of the same unique island characteristics that influence the distribution of plants and animals. But in Darwin's case the brief, three-week visit to the Galápagos was to prove to be a life-changing experience.

Oceanography

The Galápagos Islands were once termed the enchanted islands, allegedly because mariners had difficulties negotiating the complex currents swirling around the islands. One account even suggested that the islands themselves were in motion, and actually they are, but at the rate of tectonic motion, quite undetectable to sailor and ecotourist alike. It is difficult to understand the ecology of the islands without some knowledge of the pattern of the ocean currents, and the consequences of such a pattern. Virtually the first lecture every tourist group receives from its naturalist guide is an overview of the ocean currents as they affect the islands. Given that I am, in effect, the naturalist guide for this book, here goes.

Earth rotates every twenty-four hours, a motion that produces very important effects on masses of gas and liquid. In other words, both the atmosphere and the oceans are strongly influenced in their respective motions by the reality of Earth's daily spin from west to east. Because of Earth's rotational motion, surface currents in the northern oceans of the world tend to move clockwise while in the southern hemisphere surface currents move counterclockwise. (It is said that toilets flush clockwise north of the Equator and counterclockwise south of the Equator. I never remember to test this, though I know that it is not the case.)

The surface current normally of most importance to the Galápagos runs along the west coast of South America and divides and spreads gradually westward, eventually arriving at the Galápagos. This is the Humboldt Current, named for Alexander von Humboldt, the great German explorer who first described much of the ecology of South America, particularly the vegetational zonation patterns in the Andes. The Humboldt Current, coming as it does from the Antarctic, is a cold current. It divides into the Peruvian Coastal Current, which continues along the coast and eventually turns westward toward the Galápagos, the Peruvian Oceanic Current, which also angles westward toward the Galápagos, and one part of the South Equatorial Current, which swings westward south of the islands. Some of these currents become subsurface because cold water is denser than warm water and thus has a tendency to sink. But along the coastline of Chile and Peru, strong winds normally blow from the Andes, literally removing warmer surface water and permitting deeper, colder water to rise to the surface, a process called *upwelling*. This colder water helps raise chemical nutrients from the ocean depths to the surface water, providing a fertilizing effect that promotes immense growth of plankton. This planktonic abundance, in turn, supports numerous small fish (particularly anchovies) and squid, which serve as food sources for larger fish, larger squid, and ultimately a huge concentration of seabirds.

Upwelling occurs at the Galápagos as well. The Humboldt Current flows into the westward-flowing South Equatorial Current, which goes directly to the Galápagos, slowly warming as it makes its way along the Equator to the island. This surface current is joined by the Panama Flow, a warm current flowing southwestward from the Central American coast. The Panama Flow is warmed in part by the North Equatorial Countercurrent, a warm current flowing from west to east just north of the Equator. Finally there is the subsurface Cromwell Current (also often termed the Equatorial Undercurrent), a deep, cold equatorial current that flows eastward beneath the westward-flowing South Equatorial Current. When the Cromwell Current reaches the westernmost of the Galápagos Islands, the islands of Fernandina and Isabela, it upwells, bringing cold, nutrient-rich water to the surface and supporting a lush array of oceanic life. This pattern is essential to the continuing productivity of the oceans around the islands and to supporting the impressive colonies of seabirds and other wildlife. When the current patterns change and the upwelling ceases, disaster strikes the Galápagos.

El Niño, the Southern Oscillation

Most people by now are familiar with the El Niño phenomenon, but few can explain what it actually is, and no one can explain the mechanism that initiates it. El Niño, or "the child," so-named because its beginning often tends to coincide with the Christmas holidays (though it can begin as early as September), is an occasional (though increasingly common) radical alteration of atmospheric and oceanic current patterns, probably rooted initially in some sort of atmospheric change, possibly related to global warming. It really is not known how an El Niño begins, but most of the focus is now on an alteration of the intensity of the trade winds because they affect patterns of oceanic circulation. Sea surface temperatures are higher throughout the tropical Pacific Ocean, and this, when coupled with diminished trade winds, enhances the rate of evaporation of water into the atmosphere, conditions ideal for producing much rain. The alternate name for El Niño is Southern Oscillation, a term that refers to the movement of a complex system of air masses called the ITC, or intertropical convergence. The ITC is normally located north of the Galápagos and, during an El Niño, advances ever so slightly south. The effect is a shift in current patterns, which moves warmer currents such as the Panama Flow farther south, thus warming the ocean and stopping upwelling in areas ranging from coastal Peru to the Galápagos. A major El Niño can produce global changes in precipitation patterns. Its effects are by no means confined to the Pacific.

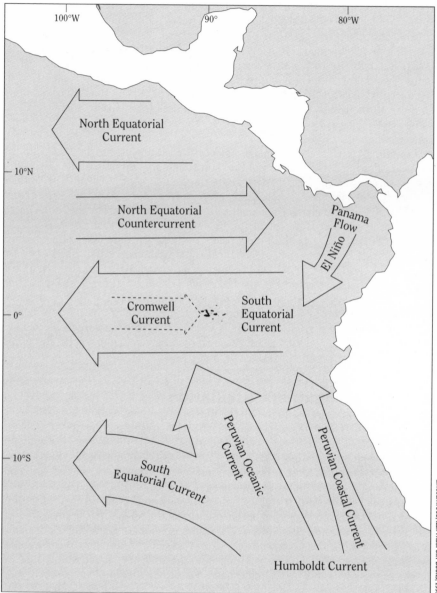

The prevalent currents that influence the Galápagos Islands. Cold water from the Humboldt Current mixes with warmer water from the South Equatorial Current. The deeper Cromwell Current brings cold, subsurface water to the islands. *Adapted from P. Constant,* Galapagos: Darwin's Islands of Evolution

During the powerful El Niño of 1997–1998, both sea and air temperatures around the Galápagos rose well above normal. The result for marine species was that the food webs dependent upon upwelling collapsed, resulting in a major mortality for species ranging from anchovies to boobies. Cold water–dependent Galápagos species such as the Galápagos penguin, marine iguana, and flightless cormorant are particularly hard hit when upwelling ceases. In May of 1997, the start of a major El Niño, I saw several abandoned sea lion pups, and there is some evidence suggesting that starvation and abandonment of pups is more common during an El Niño.

El Niño events produce anomalous rainfall patterns. Heavy rains fall in the Galápagos during El Niño years, much heavier than is typical. During the El Niño of 1997–1998, between January 1, 1997, and December 31, 1998, a grand total of 134.15 inches of rain (330.76 cm) was recorded at the Charles Darwin Research Station at Santa Cruz. Only the El Niño event of 1982–1983 recorded more rain, and that total was only 0.02 inches (0.6 mm) higher. On the island of Española, where numerous boobies nest as well as the waved albatross, the simple ground nests of these birds were partly submerged from the already intensive rainfall, and most of the eggs were doomed. Even when eggs hatched, the chicks often died of starvation because parents were not able to find enough food. It has also been suggested that the insect flush accompanying El Niños can have a devastating effect on nesting seabirds. Española in May 1997 was flush with vegetation, and mosquitoes were abundant. Large swarms of mosquitoes have been blamed for causing boobies, albatrosses, and frigatebirds to abandon nests and young.

Marine iguana carcass, Fernandina

But other species, such as the mockingbirds and Darwin's finches, often experience large population increases during El Niños. What is normally a near desert suddenly bursts into lush, junglelike habitat where food for these birds becomes abundant.

Prior to the El Niño of 1997–1998, the last significant El Niño to affect the Galápagos was one that occurred in 1982–1983. Thus it is clear that El Niños are not annual but irregularly periodic and produce short-term significant effects. Satellite imagery is now being used to track atmospheric patterns and should prove an aid in forecasting El Niño events as well as monitoring El Niños in progress.

Though plants generally prosper in the high rainfall that typifies a strong El Niño, this is not always the case. In 1999, Alan Tye and Iván Aldaz found that the highland forests of *Scalesia pedunculata* suffered major diebacks from root rot in the 1982–1983 El Niño. Fortunately these forests were not similarly affected by the 1997–1998 event. Cacti, such as the giant trees of the genus *Opuntia*, can absorb so much water in El Niño years that they topple over from their own weight, especially given the stronger winds that often occur. But though large adult cacti fell, many cactus seeds germinated from the increased moisture.

El Niño conditions can, in some cases, promote movement of species among the islands. During the 1997–1998 El Niño it is thought that the lush vegetation may have ultimately been responsible for the colonization of Fernandina and Genovesa by the smooth-billed ani *(Crotophaga ani)*, a slender bird of the cuckoo family easily identified by its black plumage and large, parrotlike bill. The species was introduced about two decades ago, reputedly to feed on ticks that were plaguing cattle on Santa Cruz. It remained rare until its population was boosted by the 1982–1983 El Niño. Apparently the lushness of the 1997–1998 El Niño had a similar effect.

Even more significantly, frogs were able to breed on the Galápagos during the 1997–1998 El Niño. Amphibians, which require moisture to maintain mucous on their skins, are poorly adapted for survival on the islands (to say nothing of getting there in the first place), but two species became established at different places on the islands. A tree frog *(Scinax quinquefasciata)*, family Hylidae, now can be found breeding around Puerto Ayora on Santa Cruz, and another frog in the large tropical family Leptodactylidae is established around Puerto Villamil on Isabela. Both of these amphibians were likely brought to the islands accidentally by human transport rather than any natural dispersal. The long-term survival probabilities for these species are dubious, but they were certainly "jump-started" by the lush conditions produced by the El Niño.

Following the cessation of an El Niño, there can be a La Niña, when air and water temperatures are cooler than normal and conditions become drier than usual. Such was the case for the Galápagos in 1998–1999. Because of the colder conditions, upwelling is enhanced, and marine life prospers, but as you might guess, terrestrial organisms may suffer from higher temperatures combined with reduced rainfall.

Climate and Seasonality

The Galápagos Archipelago, situated on the Equator, can be expected to have a tropical climate and thus be generally warm. But climate is affected by more than latitude. Ocean currents, which carry significant amounts of heat, also influence island climates by affecting local patterns of precipitation. And most importantly, because of Earth's tilt on its axis, a tilt of 23.51°, Earth is a seasonal planet. The seasons become increasingly pronounced the farther in latitude you are from the Equator. In low-latitude areas such as the Amazon Basin and the Galápagos Islands, it is normal to experience a wet and dry season in the course of a year. However, in the Galápagos, because of the influence of oceanic currents, it is customary to refer to a hot season, when it is normally but not always wet, from about December to May, and a cool season, when it is drier but not always dry, from June to November. It is during the cool season that parts of the islands are usually bathed in a mist called *garúa* (which means "misty" in Spanish), hence the term *garúa season*. Though this pattern is generally true, there is much variation from one year to another, particularly when there is an El Niño or La Niña. Years of drought and years of abundant rainfall are not uncommon. The climatic pattern is, at best, a generality.

Hot season temperatures can reach highs in the 90s (greater than 32°C). It feels truly tropical! It also rains frequently, often coming in the form of tropical downpours, but when it is not raining, skies are clear, deep blue, and magnificent. It is easy to travel among the various islands during this season because the sea is normally calm and the waters tranquil. This is because the southeast trade winds are lessened during this season.

During the hot season, the rainfall stimulates maximum plant growth throughout the islands. These are the months when the land birds reproduce, responding to the flush of insects produced by the plant growth. It is also during these months when most plants flower and produce fruit.

During the cool (garúa) season, persistent mist forms over the mountains and even over lowland areas. Caldera slopes, particularly on the southeastern sides, experience fairly steady rainfall. The garúa is caused by the cool Humboldt Current as it interacts with the warm southeastern trade winds to form a temperature inversion, where warm air rests atop cooler air. Where the two air masses meet at an elevation of about 320 to 650 feet (between 100 and 200 m), there is a strong tendency for water to condense, forming, in essence, a low-altitude stratus cloud. When the cloud reaches the islands, it is pushed up against the caldera slopes, cools the warmer air, and results in fog and mist, or the garúa.

During the garúa season the seas around the islands are much rougher, particularly during September, when the Humboldt Current is at its strongest. Remember this if planning a visit to the islands, particularly if you are prone to seasickness. It is interesting to note that when Charles Darwin visited the Galápagos in September of 1835, the HMS *Beagle* encountered strong winds when it was navigating toward Abingdon Island, resulting in a drift of 20 miles off course to the northwest. Upwelling is at its strongest during the garúa months, a fortuitous event for the nesting seabirds who raise their young during the summer months, when fish and other oceanic food is normally at its most abundant. Though the garúa can be prevalent, it does not result in much accumulated water, so in essence the garúa months represent a dry season for terrestrial vegetation and animals alike.

No climate on Earth has always been as it is today. Earth changes geologically, and it changes climatologically. Ice ages that occurred recently (in geologic time) dominated the northern hemisphere and produced significant global effects. It is possible to gain a glimmer of understanding of climate change over long periods of time by looking at what ecologists call *pollen profiles*. Pollen, the tiny male reproductive body of flowering plants, is very resistant to decomposition, especially under certain conditions. In addition, it is generally at least genus-specific, meaning that pollens from different plants can be used to identify those plants. In acidic lakes and bogs, pollen can remain unchanged for literally thousands of years. As lakes and bogs slowly but inexorably fill with sediment, they also fill with pollen. Anoxic sediments in lakes and bogs suppress bacterial growth and preserve organic material like pollen. And the deeper the sediment, the older the pollen within it. Pollen preserved from deep sediment that has required thousands of years to accumulate can be compared with that from intermediate and shallow sediments to identify the plant species that have dominated during different historical times.

In 1976, Paul A. Colinvaux and Eileen K. Schofield provided such an analysis for the Galápagos by using pollen from El Junco Lake in the highlands on San Cristóbal. Results strongly suggested that although the climate the past nine thousand years on the Galápagos has been much the same as it is today, it was considerably drier from ten thousand to thirty thousand years ago.

Such conditions would favor arid zone species such as cacti. Ecologists have noted that there are fewer endemic plant species in the moist highlands of the islands than in the arid lowlands, and they have wondered why. But if the climate was so much drier ten thousand to thirty thousand years ago, such a trend is easy to explain. Arid species have had a much longer time to be on the islands, to speciate, to become Galápagos endemics. There may well have not even been habitat suitable for highland species until quite recently. Times change.

Geology

The Galápagos have a look all their own. Everywhere there is evidence of volcanism, of islands born from forces deep beneath the seas that surround the islands. The islands that you walk upon range in age from less than a million years old to just over 4 million years old. But Earth is active, and other islands once existed, islands that are now submerged and have drifted well away from the present archipelago, some perhaps as old as 90 million years. The Galápagos are testimony to the aged and continuing interactions between the turbulent geology of the planet and the life forms that come and go as the genetic material, DNA, is forced to continually rearrange itself to accommodate the reality of change.

The Galápagos, taken together, make an excellent primer on geology, especially for those who want to major in volcanism. Vast lava fields carpet Santiago, while lava beaches characterize many of the islands. Galápagos vistas, such as those seen from Tagus Cove at Isabela or Pinnacle Rock at Bartolomé, reveal calderas and tuff cones, the work of the volcanoes (called volcáns in the Galápagos) born beneath the Pacific Ocean. Visitors to the islands soon develop a familiarity with words such as magma, fumarole, basalt, lava tube, and tephra and take pride in being able to separate "aa" from "pahoehoe" lava. Like Charles Darwin, the modern ecotourist can revel not only in the unique wildlife of the islands but also in the remarkable geology that tells such an enthralling story. No one can really look at these islands without developing some interest in just how rocks get to be rocks and how such events influence, indeed determine, the habitations of Earth's most notable characteristic—the presence of life forms.

Ecological Zones

The two dominant determinants of terrestrial ecology are temperature and precipitation. If you know the average annual temperature and average annual precipitation for any point on a continent (or island), you can pretty accurately say what sort of ecosystem will exist at that place. Of course seasonality is also an important influence, but it does not change the fundamental reality, long realized by ecologists, that mean annual temperature plus mean annual precipitation, taken together, determines if you are looking at desert, grassland, tundra, or rainforest.

Plants, being entirely sessile organisms, are extremely sensitive to small changes in climate, and plants determine the identity of terrestrial ecosystems. We do not speak of the "lion zone" but of the African savanna. We don't speak of the "blue jay forest" but of the oak-hickory forest. The diversity of the world's land plants is a straightforward reflection of the diversity of the world's terrestrial climates. It is basically climate that confines cacti to deserts and climate that restricts reindeer moss to arctic tundra. And, in an evolutionary sense, it is basically climate that made cacti and reindeer moss by the selection pressures continually exerted on plants to develop growth forms that are adapted to the realities of climate.

When Alexander von Humboldt explored the Andes, he wrote extensively of how the vegetation changed with elevation, slope, and exposure to direct sunlight. He recognized different zones of vegetation as the various plant groupings wrapped around mountains at differing elevations. And he realized that these zones, which often blended, one into another, were simply a reflection of different assemblages of plants adapted to different climates. Air cools as it rises, and so temperatures drop off as you climb up a mountain slope. As air cools, any moisture trapped in it can precipitate and form clouds and rain. So as you move up and down a mountainside, you are actually moving through several climatic regions, and you see vegetation changes that reflect the climatic gradient. Ecologists routinely describe plant assemblages as *communities,* and for a time in the history of ecology, there was a burgeoning field of study called *phytosociology,* the study of how plant communities come together.

Because the Galápagos Islands have tall volcanic cones creating a relatively sharp relief, there is an altitudinal gradient on the larger islands sufficient to produce an array of ecological zones. Precipitation varies, often dramatically depending upon whether it is a wet year or a drought year. Arid lowlands can receive as little as no rainfall or as much as 12 inches (about 30 cm). Highland areas can receive as much as 65 inches (about 170 cm) of rainfall, but they can also receive as little as 12 inches! Throughout the archipelago, variability is the rule, not the exception.

As a result of ecological zonation, on islands such as Santa Cruz or Floreana, you can move from an intertidal beach or mangrove forest to a desertlike forest of giant cacti and on up into a moist, lush forest of trees heavily clothed in epiphytes, plants that drape themselves from other plants and take in moisture from the air. Smaller islands lack altitudinal relief and thus have fewer ecological zones. The zones evident on the Galápagos are somewhat similar to those that Humboldt saw in the Andes. But there are important differences. The Galápagos have far fewer species, and some of the species have evolved into endemics, now found nowhere else but on the Galápagos. For this reason, it is fair to say that the plant ecology found on the Galápagos exists in no other place.

Plant ecologists recognize seven ecological zones on the Galápagos, though, as mentioned previously, only the larger islands, those with sufficient elevation, support all of these zones. From lowest to highest elevation, they are as follows: the coastal zone, the arid zone, the transition zone, the scalesia zone, the brown zone, the miconia zone, and the pampas or fern zone. It is important to bear in mind that these zones may be anything but sharply defined. When the land slopes gradually, the climatic gradient is also gradual, and thus one zone will slowly be replaced by another, but there will be an extensive "ecotone," or area of blending between the zones. It often takes botanical training to realize which zone is which. Zones vary in elevation with exposure to winds. In general, zones are lower on the windward than on the leeward side of an island because there is more moisture on the windward side. For example, on Santa Cruz, the moist scalesia zone drops to about 650 feet (about 200 m) on the windward side of the island but is not encountered until an elevation of about 1,900 feet (about 570 m) on the leeward side. Nonetheless, the zonation concept has utility for describing the basic plant ecology of the islands, and so I'll outline it here.

Every visitor to the islands usually sees at least five or six of the seven zones: coastal, arid, transition, scalesia, miconia, and pampas. A ride along the road that traverses Santa Cruz, from Puerto Ayora up into the highlands and back, will cross each of these zones. The brown zone is hard to find because it is transitional between the scalesia and miconia zones and because it is an area that has been extensively converted to agriculture.

The Coastal Zone

The coastal zone is, of course, present on all of the islands. It may support a mangrove forest or consist of dense shrubs or scattered plants above the tide line. It is easy to recognize some of the most common coastal zone plant species. Look for saltbush *(Cryptocarpus pyriformis)* pretty much anywhere along the shorelines. Some islands support dense stands of this salt-tolerant shrub. The small, curvy, unlobed evergreen leaves are thick, an adaptation to retain moisture. Winds blow strongly along coastal areas, and saltbush, like other species described below, is adapted to retard water loss to evaporation. Do not confuse saltbush with salt sage *(Atriplex peruviana)*. Salt sage can grow as a small shrub but is more commonly seen as a short mat. Leaves are small and wrinkled, and little clusters of yellow flowers can sometimes be found on the branch tips. Two species in the genus *Sesuvium* are found throughout the archipelago. There is no common name for these plants. The tiny island of Plaza Sur abounds in these *Sesuvium* species, much to the delight of the resident land iguanas. The plants, which grow in dense, prostrate mats, have very thick, small leaves that are facultatively photosynthetic. During the dry times the leaves become reddened because chlorophyll is not manufactured and auxiliary pigments become increasingly apparent. Once it rains, of course, the plants rapidly synthesize chlorophyll, and the leaves turn bright green. Blossoms are either white or violet, depending upon species. Beach morning glory *(Ipomoea pes-caprae)* is unmistakable, a winding, prostrate vine with prominent, colorful, trumpet-shaped flowers. It is common in sandy areas throughout the islands.

Mangroves are tropical coastal trees that are highly tolerant of salt, trees that ecologists like to call *halophytes*. The word *mangrove* is generic, not taxonomic. Many kinds of mangroves exist in the tropical world, and they represent many families of plants. Some, like the red mangrove *(Rhizophora mangle),* are basically cosmopolitan in equatorial regions. This is because they have extraordinary dispersal powers. Red mangroves are common on many coastal areas of the Galápagos, but three other mangrove species occur as well: black mangrove *(Avicennia germinans),* white mangrove *(Laguncularia racemosa),* and button mangrove *(Conocarpus erectus).*

Mangroves are typically short in stature, often not more than about 10–20 feet (3–6 m). Leaves are thick and waxy, typical of plants adapted to retard water loss to evaporation. Red mangroves are easy to recognize by their mazelike array of reddish prop roots that dangle down from the branches, some of the roots penetrating the substrate and anchoring the plant, some just growing toward that end. An established stand of red mangroves, such as can be seen at Black Turtle Lagoon on northern Santa Cruz or at the Charles Darwin Research Station at Puerto Ayora, forms habitat for many species of marine organisms as well as birds. Just look at the prop roots, typically coated with various sessile marine creatures ranging from oysters to sea anemones. Red mangrove seedlings are contained in long, pencillike pods that float horizontally in the sea and can travel remarkable distances in the currents. When in shallow water, they reorient vertically in the water column, and when they touch bottom along a beach or mudflat, they can quickly anchor and take root. Black mangroves are most notable for their myriad upward-poking roots, like masses of pencils sticking out of the muddy ground. These roots, called *pneumatophores,* act to aid the plant in taking in oxygen from air because the plant grows in anoxic mud.

The Arid Zone

Most visitors to the islands spend more of their time wandering through the arid zone than any other zone. There are many unique plants to be found here, including some of the most characteristic of the islands.

One species of tree, the palo santo *(Bursera graveolens)* dominates this zone. It is a deciduous tree, leafing out during times of high moisture, then dropping its leaves and looking gaunt and lifeless during dry times. The sight of a barren palo santo forest against the volcanic backdrop of the islands does, indeed, add to the starkness, the barrenness, that so many mariners (and Darwin) felt when first seeing the Galápagos. These trees emit a distinct odor, giving them their popular name of "holy tree." Palo santos grow widely spaced, almost orchardlike, covering hillsides at low elevations throughout the islands. When in leaf they have delicate, compound leaves with seven leaflets. Flowers are whitish and emit a fragrant odor. The bark is dark but is typically covered by pale gray lichens, giving the trees a pale, skeletal look during dry periods.

Aside from the abundant palo santo trees, it is the cacti that next claim your attention in the arid zone. Two kinds grow to the size of respectable trees: the prickly pear (a species of the genus *Opuntia*) and the candelabra (of the genus *Jasminocereus*). They are easy to distinguish because prickly pears are shaped like more conventional trees, though with wide, rounded, spine-covered pads (called *cladodes*) instead of leaves, and candelabras are multistemmed, the stems looking a bit like prickly green sausage links. Prickly pears can grow to 40 feet (about 12 m), while candelabras are shorter, reaching about 22 feet (about 7 m). Both of these kinds of cacti abound in places like the Charles Darwin Research Station, but they can be seen at many other places throughout the islands. There is a splendid forest of prickly pear cacti on Santa Fé, as well as some dandy examples on Plaza Sur.

Prickly pears colonized the Galápagos from the Americas, where there are in excess of three hundred species. There are six species found on the Galápagos and thirteen varieties within those species. They are given scientific names such as *Opuntia echios* var. *gigantean,* the largest prickly pear of the lot. My guess is that these cacti have long been residents of the islands. There are two reasons to believe this. One is their stature; they have grown literally to be trees, something that no mainland prickly pears have accomplished. This result of evolution speaks to a lack of competing tree species. The second is that so many Galápagos animals are dependent on prickly pears, making the prickly pear what ecologists call a keystone species. Take away the prickly pears, and you begin a domino effect that would likely be catastrophic to animals such as cactus finches, mockingbirds, tortoises, and land iguanas. Indeed, there might have been an evolutionary "arms race" of sorts that occurred between the prickly pears and some of the tortoises: The plants grew taller,

Tree-sized *Opuntia* species, Santa Fé

but the tortoises grew longer necks and a notch in their shells to enable them to better reach upward to obtain the prickly pear pads.

There is yet another cactus that seems to grow directly from volcanic rocks, the lava cactus *(Brachycereus nesioticus)*. You can find this remarkable plant on places such as Punta Espinosa on Fernandina or on the lava fields of Santiago. It is not treelike but matlike, growing in dense little clumps, its wide rounded stems utterly coated with spines. It is amazing how this plant has adapted to be the pioneer invading species on lava flows.

There are four endemic species of tiny spindly plants in the genus *Tiquilia*. Like the lava cactus, *Tiquilia* species also thrive where others fear to root—on bare, exposed lava soils. Visitors who walk the many steps up the tuff cone on Bartolomé see the widely scattered species of *Tiquilia*, sometimes wondering, between huffing and puffing up the slope, what those curious pale plants may be.

The arid zone encompasses some lush areas, and some plants with a distinctly tropical affinity thrive in these areas. The passionflower *(Passiflora foetida)* is one of a large number of species that are found throughout much of the American tropics. Only three passionflower species occur on the Galápagos, and only the one listed here is common. These vines are known for their complex and colorful flowers that are pollinated by butterflies. Mainland species are well-known to have leaves that contain concentrated cyanide compounds that retard herbivores such as caterpillars. Another plant with tropical affinities is lantana *(Lantana peduncularis)*. This plant grows as a shrub and has small white vaselike flowers.

Manzanillo, the poison apple tree *(Hippomane mancinella)*, is common in many areas in the arid zone. It produces green applelike fruits, but though you are on Darwin's Eden and not the other one, don't be tempted because the fruits, like the sap, are poisonous.

It is possible to trip should your feet get caught up in goat's head, a dense little prostrate vine in the genus *Tribulus*. Its seeds are spiked and very sharp (be careful handling them) and said to resemble a goat's head, though I think they look more like a medieval mace. The vine itself is easy to recognize from its small, usually hairy, grayish compound leaves. It is very common in many areas, and the seeds are eaten by Darwin's finches.

Should you be wandering along in the arid zone and see a tomato, you have found the Galápagos tomato *(Lycopersicon cheesmanii)*. This species is endemic to the islands. It has a conspicuous yellow flower.

But a plant with an even more obvious yellow flower is *Cordia lutea*, usually a small shrub, sometimes a tree. Popularly called *muyuyo*, this plant bears large trumpet-shaped blossoms that are usually fairly fragrant. The plant's bark is grayish, but its leaves, in addition to its flowers, make it easy to identify. Darker green than the leaves of most other plants of the zone, the leaves of *Cordia lutea* are covered with tiny dense hairs and are very rough, almost like sandpaper. The plant is common on most islands in the arid zone.

Looking at botany from the standpoint of evolutionary biology, probably the single most important genus of plants on the Galápagos is the genus *Scalesia*. It can be said that these plants are the botanical equivalent to Darwin's finches. Like the famous finches, the considerably less famous *Scalesia* species have evolved extensively and have diversified into fourteen species as well as an additional half-dozen subspecies and varieties. The scalesia zone is named after *S. pedunculata,* an abundant species, but the various species of *Scalesia* range throughout the arid, transition, and scalesia zones. Different islands have different species, a parallel of sorts to the finches' distribution. The local name for *Scalesia* on the islands is *lechoso,* and there is certainly a lot of diversity among the various lechosos. Leaves show a remarkable divergence in shapes and sizes, and it is beyond the scope of the treatment here to attempt to detail the differences among species. (See the references at the end of this chapter for guides to plant identification.) All lechosos are in the huge plant family Compositae, the family that contains sunflowers and daisies, so they all have daisy-like flowers. But otherwise the lechosos show such amazing differences that it is unlikely a casual observer would know them to be members of the same genus.

The Transition Zone

In order to move from a region of aridness to one of lushness, it is necessary to make a transition. Ecologists recognize this and have named the zone between hot/dry and cool/wet as the transition zone. This zone can be tricky to find on the Galápagos because its distinguishing characteristics can be subtle. Lots of arid zone species, such as palo santo, make it into the transition zone, and lots of scalesia zone species (the next highest zone after the transition zone) do so as well.

Ecologists use indicator species to help them recognize when they are within a certain ecological designation. For example, candelabra cactus is certainly an indicator species of the arid zone. The best indicator species of the transition zone is a small tree with brownish hairlike "wool" beneath the leaves, the pega pega *(Pisonia floribunda)*. Pega pega is a Galápagos endemic usually found most abundantly between 300 and 600 feet (90 and180 m), the typical elevation of the transition zone (though it may be higher on some islands). An additional common species in this zone is guayabillo *(Psidium galapageium),* another endemic small tree (or shrub), this one with grayish bark and very dry, leathery leaves. The combination of both guayabillo and pega pega pretty much defines the transition zone. It is interesting to compare these two species because their anatomies, at least regarding their leaves, suggest actual transition from arid to more moist areas. Though they are evergreens, as would typify a more moisture-laden region, the plants' leaves still retain adaptations that retard dessication: woolly leaves (pega pega) and leathery leaves (guayabillo).

One other characteristic of the transition zone, often observed on the trees just described, is that they are usually laden with epiphytes, lichens that become increasingly abundant as moisture increases. This characteristic is further enhanced in the next highest zone.

The Scalesia Zone

I have traveled fairly extensively in the Andes, and the scalesia zone of the Galápagos never fails to remind me of some of the elfinlike forests of the high Andes: cool, damp, often shrouded in an ethereal fog, the stunted trees draped liberally with wispy epiphytes. On the Galápagos the scalesia zone (and higher zones) is bathed in the garúa mist for much of the year. The zone is dominated by a single tree species *(Scalesia pedunculata),* which, like other scalesias, is also known as a lechoso.

Scalesia pedunculata is particularly abundant between 650 and 1,300 feet elevation (about 200–400 m), essentially defining the scalesia zone. The tree, which rarely exceeds 50 feet in height (about 15 m), is spindly and multibranched, with a canopy shaped roughly like an umbrella.

Epiphytes, sometimes popularly called *air plants,* are plants that grow on plants. They use their host only as a perch and thus are not parasitic. They do not invade and take nutrients but instead only occupy surface area on a bole or branch. Epiphytes are typically abundant from lowland tropical rainforests through moist Andean forests. On the Galápagos, the lechoso tree is typically a host for many epiphytes, including one species of bromeliad *(Tillandsia insularis)*. Bromeliads, of which there are about two thousand species, are members of the pineapple family, and many (but not pineapples) are epiphytic, their spiky leaves arranged cisternlike so that they trap rainfall. In tropical America, where bromeliads are abundant, many hummingbird species help cross-pollinate the colorful red or violet flowers that grow on a spike extending beyond the cuplike leaves. It is the large carpenter bee *(Xylocopa darwini)* that performs this function on the Galápagos, where no hummingbirds live.

Fairly uniform stands of S. *pedunculata* are typical, though many other species can be found as shrubs and herbs in the understory. Arid zone species, such as palo santo, are also common in parts of the scalesia zone. Because of the ample precipitation found in the scalesia zone, it is one of the best regions for agriculture, and on some of the major islands, such as San Cristóbal and Santa Cruz, the natural vegetation has given way to agroecosystems.

The Brown Zone

Ecologists identify the brown zone based on the abundance of cat's claw *(Zanthoxylum fagara)*, a small shrubby tree with sharp spines that give the plant its common name. But besides cat's claw, this is a zone where epiphytes become particularly abundant, especially liverworts, evolutionarily ancient plants. The bark and branches of the scalesias and cat's claws are typically covered with liverwort masses, as well as various mosses. The ground is soft, like uneven thick pile carpet, because mosses and various ferns grow in profusion. It is possible here to see the tree fern *(Cyathea weatherbyana)*, which, as the name implies, grows to the size of a small tree. Tree ferns are common throughout the Andes. The name *brown zone* comes from the accumulation of fallen epiphytes that litter the ground.

The Miconia Zone

Miconia is a common shrub throughout the Latin American tropics and often grows in disturbed areas. There are numerous species. On the Galápagos, there is but one species, an endemic called the cacaotillo *(Miconia robinsoniana)*. It dominates at upper elevations on two of the major islands. Miconia leaves are oval with very pronounced veins easy to identify. The miconia zone is essentially above the tree line, usually around 3,300 feet (about 1,000 m). Here the landscape changes from small dense forest to open shrub land with a dense carpet of ferns interspersed with abundant miconias. If you want to see the miconia zone, you must visit Santa Cruz or San Cristóbal because these islands are the only places where the endemic *M. robinsoniana* can be found.

The Pampas Zone

Pampas are grassland ecosystems that can be found in Patagonia, Argentina, and Chile. On the Galápagos, pampas are not the same kind of ecosystems. On the large islands, the windswept pampas zone, also called the *fern zone,* is found where exposure is greatest and precipitation is most abundant. Total annual rainfall can approach 100 inches (about 250 cm) in this zone, making it well within the range of precipitation in a tropical rainforest. Like the miconia zone, this zone is well above the tree line and consists mostly of ferns, especially bracken fern *(Pteridium aquilinum)*. Also common are sedges, grasses, and various mosses. The pampas zone is reminiscent of heath lands, the ground soft and boggy, very wet. Peat bogs are frequently encountered. Tree ferns also grow in this zone, and where they occur, these 12-foot-tall plants are the tallest form of vegetation.

Selected References

Berry, R. J., ed. 1984. *Evolution in the Galápagos Islands.* New York: Academic Press.

Carlquist, S. J. 1965. *Island Life.* Garden City, N.Y.: Natural History Press.

Colinvaux, P. A. 1972. Climate and the Galápagos Islands. *Nature* 240:17–20.

Colinvaux, P. A. and E. K. Schofield. 1976. Historical ecology in the Galápagos Islands: I. A Holocene pollen record from El Junco Lake, Isla San Cristóbal. *Ecology* 64:989–1012.

Cox, C. B. and P. D. Moore. 1993. *Biogeography: An Ecological and Evolutionary Approach.* 5th ed. Oxford: Blackwell.

Jackson, M. H. 1993. *Galápagos: A Natural History.* Calgary, Alberta, Canada: University of Calgary Press.

Johnson, M. P. and P. H. Raven. 1973. Species number and endemism: The Galápagos Archipelago revisited. *Science* 179:893–95.

McMullen, C. K. 1999. *Flowering Plants of the Galapagos.* Ithaca: Cornell University Press.

Porter, D. M. 1976. Geography and dispersal of Galápagos Islands vascular plants. *Nature* 264:745–46.

———. 1984. Relationships of the Galápagos flora. *Biological Journal of the Linnean Society* 21:243–51.

Schofield, E. K. 1984. *Plants of the Galápagos Islands.* New York: Universe.

Schonitzer, K. 1975. *Galápagos Plants.* Publication No. 172. Quito, Ecuador: Charles Darwin Foundation.

Snow, D. W. and J. B. Nelson. 1984. Evolution and adaptations of Galápagos Sea-birds. *Biological Journal of the Linnean Society* 21:137–55.

Thornton, I. 1971. *Darwin's Islands.* New York: Natural History Press.

Tye, A. and I. Aldaz. 1999. Effects of the 1997–98 El Niño event on the vegetation of Galápagos. *Noticias de Galápagos* 60:22–24.

Wiggins, I. L. and D. M. Porter. 1971. *Flora of the Galápagos Islands.* Palo Alto, Calif.: Stanford University Press.

4

Eminently Curious:
Charles Darwin and the Galápagos

C harles Darwin and Abraham Lincoln both entered the world on the same day, February 12, 1809. The future president of the United States was born in much humbler circumstances than the future author of *On the Origin of Species*. Charles Darwin's father, Robert Darwin, was a highly successful physician, and the family owned a large, comfortable house in Shrewsbury in central England. Charles had one younger brother and three older sisters. But it was his father who loomed over him, whom he wanted so much to please, and whom he had to convince to permit him to make the voyage on the *Beagle*. His father told Charles that he could go if he, just graduated from Cambridge and his prospects limited, could find but one person whom he respected who would urge that Charles be allowed to join the ship.

Luckily for Charles, his uncle Josiah Wedgewood, whose daughter Emma would later become Darwin's wife, was happy to act on his behalf. Otherwise, Darwin might have become a country parson, he would never have gone to the Galápagos, and history would be, shall we say, different. It has been suggested more than once that Uncle Jos may have wanted his nephew to gain some worldly experience before becoming too amorous with his daughter. Be that as it may, it was Uncle Jos, along with Charles's Cambridge mentor, John Henslow, who made Darwin's eventual fame possible. Henslow was originally offered the *Beagle* position but could not accept it and recommended his star student Darwin instead.

Charles had to pass a test first. He had to be interviewed and accepted by Captain Robert FitzRoy, the youngest captain in the Admiralty, an aristocratic young man Darwin's age who was looking for someone of equal social stature with whom to converse during the long voyage.

Although Darwin would eventually be given the title of "naturalist," this was not the main reason for his being invited to go on the journey. FitzRoy needed company, and Darwin was. Although Darwin and FitzRoy hit it off rather well, Darwin's nose almost earned him a rejection. FitzRoy believed, as many did at the time, that the conformity of one's facial features said much about one's character. Darwin's nose,

Charles Darwin

rounded and gentle, did not indicate to FitzRoy that Darwin had sufficient resolve for such an arduous feat as a long voyage at sea (FitzRoy's nose was sharp and well defined). Darwin mentioned this incident in his autobiography, stating that after FitzRoy got to know him, he was "afterwards well-satisfied that my nose had spoken falsely."

"You care for nothing but shooting, dogs, and rat-catching, and you will be a disgrace to yourself and all your family." In his autobiography, written largely for his grown children and completed in 1876, six years before he died at the age of 73, Charles Darwin recalled his father's description of him as a child. Darwin then commented, "But my father, who was the kindest man I ever knew, and whose memory I love with all my heart, must have been angry and somewhat unjust when he used such words." The elderly Charles Darwin was reflective throughout his autobiography. His life had been satisfying, and his mind was at peace. Even his vivid memory of a father's cruel words did not elicit other than a bemused dismissal. Darwin knew he had been a disgrace neither to himself nor to his family.

But at one place in his autobiography Darwin did show outright contempt, a contempt so harsh as to be edited out of the first published editions of the work but, fortunately for history, restored in later editions. Darwin made very blunt remarks about Christianity, calling it, at one point, "a damnable doctrine." His abuse by those who believed him to represent a threat to Christian religious beliefs was finally answered in tones as harsh as any ever directed at him. But, despite his anger, Darwin was introspective and seemed somewhat surprised by his loss of faith. He wrote how even he was amazed as skepticism crept over him as he matured, how he was once a firm believer in the literal truth of the Bible, and how he tried to hold on to his beliefs even in the face of mounting, later overwhelming, evidence to the contrary. He was suggesting, of course, that he did not come to his belief in evolution casually or with any real degree of speed. Rather it sounds more like he grudgingly yielded to evolution and tried, at least for a while, eventually without success, to hold on to a paradigm, special creation, that had satisfied him but that at one very important point in his life, to put it in the vernacular of today, no longer fit the data. Just when and how did Charles Darwin come to believe in evolution? Exactly how did the visit he made to the Galápagos Islands contribute to the paradigm change that shook the intellectual world?

In his *Journal of Researches into the Natural History and Geology of the Countries Visited during the Voyage of the H.M.S.* Beagle *round the World,* first published in 1839, and now retitled simply *The Voyage of the* Beagle, Darwin, often prone to understatement, began his discussion of the animal and plant life of the Galápagos: "The natural history of these islands is eminently curious, and well deserves attention." He gave it some attention but perhaps not as much as you might think, given how important the islands are to the story of evolutionary biology. In *The Voyage of the* Beagle there is a full chapter on the islands, one of twenty-one chapters in the book. The pages devoted to the Galápagos represent about 6 percent of the total, not very much for the place that allegedly spawned Darwin's conversion from creationism to evolution.

But Darwin is best known for his monumental work *On the Origin of Species,* which was first published on November 24, 1859, and went through six editions. *On the Origin of Species* remains arguably the most important book ever written about biology and, given the philosophical significance accorded the elucidation of the first mechanistic theory for evolution (natural selection), one of the most important in Western civilization. It is in this work that Darwin articulated in meticulous detail his theory of evolution through common descent. It is here that he presented the theory of natural selection, how the struggle for existence within any population results in a nonrandom survival of the fittest, those organisms that survive to reproduce being those that are best adapted to the current circumstances. *On the Origin of Species* is Darwin's case for evolution, his "one long argument." But in *On the Origin of Species,* Darwin related even less about the Galápagos than he did in *The Voyage of the* Beagle. All told, only about 1.1 percent of the text of *On the Origin of Species* discusses the Galápagos. Far more words are devoted to a protracted discussion of the domestication of pigeons!

Darwin, in *On the Origin of Species,* made reference to the Galápagos in his chapter on geographical distribution on oceanic islands. He noted the large number of endemic species typically found in the bird communities of such places: "Thus in the Galápagos Islands nearly every land-bird, but only two out of the eleven marine birds, are peculiar; and it is obvious that marine birds could arrive at these islands more easily than land-birds." Darwin continued, making an argument for the unique nature of the Galápagos vegetation: "In the plants of the Galápagos Islands, Dr. Hooker has shown that the proportional numbers of the different orders are very different from what they are elsewhere." Darwin's main reference to the Galápagos was clearly intended to persuade his audience that island species bore literal affinities (implied as being of a hereditary nature) to species present on the nearest mainland, though island species were nonetheless recognizably distinct (that is, they had evolved). He chose the Galápagos as the single best example to make his case:

> I will give only one, that of the Galápagos Archipelago, situated under the Equator, between 500 and 600 miles from the shores of South America. Here almost every product of the land and water bears the unmistakable stamp of the American continent. There are twenty-six land birds, and twenty-five of these are ranked by Mr. Gould as distinct species, supposed to have been created here; yet the close affinity of most of these birds to American species in every character, in their habits, gestures, and tones of voice, was manifest. So it is with other animals, and with nearly all the plants.

Note how Darwin used the words "supposed to have been created here" in the context of a circumstantial but nonetheless logical counterargument: that these "creations" bore a nontrivial similarity to mainland species, and thus he thought it reasonable to believe that they evolved from such species. Finally, Darwin used the Galápagos as an example of how, within an archipelago, there is divergence among closely related species:

> Thus the several islands of the Galápagos Archipelago are tenanted, as I have elsewhere shown, in a quite marvellous manner, by very closely related species; so that the inhabitants of each separate island, though mostly distinct, are related in an incomparably closer degree to each other than to the inhabitants of any other part of the world.

Clearly Darwin used the Galápagos for two prongs of his argument: that the flora and fauna of the archipelago are uncommonly related to mainland species, the *nearest* mainland species, and that among the islands themselves species diverge from one another.

Remarkably, however, Darwin wrote nothing in *On the Origin of Species* about the Galápagos finches, the very group that supposedly supplied the "smoking gun" as evidence to bolster his case for evolution. Instead he referred to the mockingbirds, which he called *mocking-thrushes:* "In the Galápagos Archipelago, many even of the birds, though so well adapted for flying from island to island, are distinct on each; thus there are three closely-allied species of mocking-thrush, each confined to its own island." It would seem that by that point he had at least set the stage for a finch discussion, but Darwin asserted nothing. What a surprise, given the superb model of evolution that the Galápagos finches seemingly represent. Rather than being the very keystone example of evolution, rather than being prominently discussed throughout *On the Origin of Species,* they are most conspicuous by their absence.

And that's not all that is missing. There is no discussion of the giant tortoises, even though while on the islands Darwin became aware of how these large reptiles varied from one island to another. If the Galápagos, as is often stated in the popular literature, are a "laboratory of evolution," then why did Charles Darwin make so little use of the data from that laboratory when he assembled his case? What, exactly, was Charles Darwin's intellectual relationship with the Galápagos Islands? How much of the story of Darwin's conversion to evolution can be attributed to his experiences on the islands?

Myths are part of our culture, and Darwin has certainly become part of a commonly promulgated myth. Some college textbooks, naive nature films, and popular writings about biology tend to present a picture of Darwin on the Galápagos not unlike the story of Isaac Newton and the famous apple tree. Newton allegedly observed an apple drop from a tree and quickly converted that simple observation into the universal theory of gravitation. In the case of Newton, the story might very well not be a myth at all. But not in Darwin's case. Darwin, so the myth would have us believe, spent a few days on a remote volcanic archipelago abounding in odd birds and reptiles, experienced a sudden and dramatic intellectual metamorphosis, and realized that these creatures must have evolved and not been separately created. The myth portrays him, tall and lean in his youth, walking across the black volcanic sands of the arid islands, deep in thought, hands clasped behind his back, his mind in turmoil.

Some years ago I saw a film that contains this very scene. In it Darwin sees the obvious adaptive differences in the finches, particularly their beaks. Some have large beaks, capable of cracking the hardest of seeds. Others have medium beaks, and still others small beaks. Some have beaks so slender as to resemble beaks of insectivorous birds, not those of finches. A cluster of little birds, clearly different, yet so much alike. Darwin also realizes that the mocking-thrushes differ in subtle but distinct ways from island to island, as do the giant tortoises. He ponders this extremely odd representation of life forms and takes note of how similar some Galápagos plants and animals are to some on mainland South America. Similar . . . yet different. It all begins to slowly but inevitably add up. No logical Creator would make these life forms as they are and so curiously distribute them among these islands. The pattern suggests otherwise; it suggests that they evolved from ancestors found on mainland South America, ancestors that long ago accidentally populated the remote Galápagos and evolved into the strange and wonderful forms now found there. Such is Darwin's alleged Galápagos epiphany. As he boards the *Beagle* after experiencing the Galápagos, thoughts of transmutation of species creep from the innermost recesses of his fertile mind. He willingly stops trying to repress these revolutionary, supposedly blasphemous ideas. His mind now in intellectual agitation, Darwin cannot wait to return to England so that he may fully develop a theory about how evolution might occur, and he eventually discovers the concept of natural selection. The finches of the Galápagos become the model for this process.

The myth of Darwin and the Galápagos is understandable up to a point. Though Darwin did not devote the lion's share of *The Voyage of the* Beagle to the Galápagos, he did, in the Galápagos chapter, make isolated statements that sound like he is leaning strongly in favor of evolution: "Hence, both in space and time, we seem to be brought somewhat near to that great fact—that mystery of mysteries—the first appearance of new beings on this earth." He sounds like he was standing before a great sea of evolution and was cautiously putting a toe in the water. Soon he would swim.

This scene is, of course, not accurate. Darwin did not become an evolutionist while on the Galápagos, nor even during the *Beagle* voyage. It was not until he was safely back in England and began the serious work of compiling and interpreting his numerous specimens that he became an evolutionist. It is true that Darwin made observations on the Galápagos and elsewhere that may have made him spend some of his shipboard nights pondering "the species question." In Patagonia he found skeletal remains of extinct ground sloths that resembled modern tree sloths, as well as extinct glyptodonts that resembled modern armadillos. Why did extinct forms resemble extant forms, and why were both found in South America and nowhere else in the world? Darwin, who was a close friend of the eminent British geologist Charles Lyell, was himself a skilled geologist, a firm believer in Lyell's doctrine of uniformitarianism. Lyell purported that many landscapes were once seascapes and vice versa and that Earth is far older than traditional accounts based on the Old Testament would have you believe, that Earth not only is ancient but is and always has been changeable. Darwin had no difficulty whatsoever with such notions; indeed, they seemed patently obvious. But what effects would such changes bring upon life forms? Perhaps Darwin wondered about this, but to what degree?

It was not until he had returned to his native England and consulted with a prominent ornithologist named John Gould that he fully embraced the truth of evolution. Further, it was not until his close friend and colleague Joseph Hooker had made a detailed study of the Galápagos plants that Darwin gained his full measure of confidence in just how unique these islands really are in terms of their natural history, and what is implied by such a unique pattern. Hooker published his study in 1846. Its full title was "An Enumeration of the Plants of the Galápagos Archipelago, with Descriptions of Those Which Are New."

The Galápagos myth makes science appear to happen by sudden, quantum leaps in thinking. A few finches, some odd tortoise shells, and bingo—creationism is wrong; these animals must have evolved. But the human mind is, of course, far more complex than such a simple model would suggest. The thought process is labyrinthine, rarely direct. It usually yields to radical ideas only grudgingly, after lots of dead ends. Darwin, who came from a family of freethinkers (Charles's grandfather, Erasmus Darwin, wrote zoological poetry and speculated freely about the possibility of evolution), was himself a gifted thinker whose mind was open to what he was seeing. He was far above average in his ability to assimilate information and derive original conclusions, and he was, to borrow from his own description of the Galápagos flora and fauna, "eminently curious." The real story behind Darwin's conversion to evolution is far more interesting than that contained in the Galápagos myth outlined above, and it bears repeating. It provides valuable insight into Darwin as a thinker and scientific discovery as a process.

How do we know the Galápagos myth is a myth? How do we know the stages of thinking Darwin passed through as he gave up creationism in favor of evolution? How do we know when Darwin actually became an evolutionist?

Fortunately for historians of science, Charles Darwin left a great abundance of personal notes, journals, letters, and other material. His voluminous correspondence is largely intact. His most private notebooks are available to historians. Indeed, these notebooks have been reprinted and are easily obtained by the general reader, complete with footnotes by Darwinian scholars. Even his grocery lists and diaries of personal medical problems, including his daily levels of flatulence, are now in the public domain. This vast wealth of information has, over the past several decades, spawned a kind of cottage industry in Darwiniana. More recently, much more accurate and insightful biographies of Darwin have appeared, and many books have been published about various aspects of his life and interests. Books that he authored have been reprinted and have undergone a renaissance of new readership. Historical journals abound with scholarly papers about Darwin. What has emerged from the collective efforts of Darwin scholars is a view of the man rather different from the one entertained for the first one hundred years or so after *On the Origin of Species* was published.

Beginning in about 1959, the centennial year of the publication of *On the Origin of Species,* a newer, much more interesting picture of Darwin emerged. For an analysis of Darwin's complex thought processes that led to his conversion to becoming an evolutionist, as well as for a clear understanding of Darwin's use and nonuse of the Galápagos examples, we are indebted to Frank J. Sulloway of Harvard University. Using the wealth of Darwin material now available, Sulloway has expertly unraveled the complexity of Darwin's relationship with the islands. Much of the account included here is taken from Sulloway's various publications.

Darwin visited the Galápagos for about a month in 1835, exploring, to various degrees, four islands: Chatham (San Cristóbal), Charles (Floreana), Albemarle (Isabela), and James (Santiago). He collected giant tortoises, mockingbirds, finches, and other examples of Galápagos fauna. And he collected plants. Ironically, it would be the plants that would prove most important to Darwin's confidence about how the Galápagos represent a collective example of evolution. And all because of a few observational mistakes and some sloppy collecting on his part.

Darwin made two errors relating to the finches, one an error of omission, one an error of commission. He erred in not carefully noting from which islands the various finches were collected. And he mistakenly identified the cactus finch as a kind of blackbird and the warbler finch as a kind of wren—he did not realize at the time that both of these birds are, indeed, finches. Why were these errors important?

When Darwin returned to England, he assigned the *Beagle* specimens to various experts for classification and analysis. The result of all of this collective effort would be eventually published as a five-volume tome, *Zoology of the Voyage of the H.M.S. Beagle* (1838–1841). England's foremost ornithologist at the time was John Gould, and it fell to him to have a close look at the birds of the *Beagle* expedition. Darwin's first experience with Gould was pleasing. Darwin had brought back a skin of a small rhea (a flightless, ostrichlike bird), collected in the pampas of South America, which he correctly recognized to be different from the larger rhea species common throughout much of Patagonia. Gould agreed that it was a species new to science and named it *Rhea darwinii*, Darwin's rhea.

Gould met with Darwin to discuss the Galápagos finches sometime between March 7 and 12, 1837, about five months after Darwin's return to England and about a year and a half after Charles left the Galápagos. As Sulloway pointed out, Darwin's notes from his visit with Gould demonstrate that Darwin left this meeting with a clear realization of just how unusual and potentially important the Galápagos birds were. If Darwin ever really did experience an epiphany about evolution, it was clearly at the meeting with John Gould. It's hard to say if it really was an epiphany, given that Darwin may have had a "mind prepared" to receive the import of what Gould told him. He had clearly thought much about the distribution of organisms and had, to a degree at least, pondered the meaning of species.

Gould did not tell him that the finches evolved; Darwin concluded that for himself after the meeting. What Gould told Darwin was that there were actually thirteen species of finches and that the cactus finch and warbler (or "wren") finch were, indeed, finches. It was Gould who told Darwin that the Galápagos finches were unique to the Galápagos, an entirely new group of birds, all closely related. It was Gould who told Darwin that the Galápagos avifauna as a whole, while unique in its own right, nonetheless bore an unmistakable similarity to that of mainland South America. Darwin had not ventured north of Lima, Peru, and had no direct knowledge of the birds of equatorial western South America. It was Gould who told Darwin that the Galápagos hawk, the bird Darwin thought to be a caracara, a kind of carrion-feeding hawk, was, in fact, a kind of soaring hawk, a buteo. And it was Gould who pointed out that the various mockingbird species geographically replaced one another from island to island, with different species on different islands.

When Darwin left the meeting with Gould, he must have been, to use Sulloway's word, "stunned" at what he had learned. It would be very difficult to continue to believe that species must be immutable, given the reality of the Galápagos birds. But Sulloway made a second point: that Darwin must have been equally impressed by how difficult it was to know what, in fact, really is a species. Clearly he had been mistaken about the cactus and warbler finches. Certainly he had been confused, thinking some of the ground finches to be mere varieties when here was Gould telling him that they were separate species. The definition of species was fuzzy, with experts frequently disagreeing. Darwin himself pondered several definitions of species as he considered the matter. How could species be created immutably and yet encompass so much ambiguity?

Darwin would later use this very point in *On the Origin of Species* to convince his readers, somewhat slyly, that it is best left to experts to decide whether a particular population should be designated as a variety, a race, or a species. If evolution is true, it would indeed follow that species status would on occasion be hard to determine with any degree of exactitude. After all, the presumption is that evolution is ongoing, so there will always be numerous "gray areas" within various taxa. So it must have seemed to Darwin. Indeed, throughout the six editions of *On the Origin of Species,* Darwin never clearly defined what a species is.

Thus it was, after all, the Galápagos finches that, thanks to Gould's analysis of them, made Darwin into an evolutionist. And he became one sometime in March of 1837, more than two decades before *On the Origin of Species* would see the light of publication. Soon after his meeting with Gould he began notebooks on transmutation of species. Naturally you would expect to go to Darwin's notebooks and read lots about the finches. But there is not a word. Why?

The answer, according to Sulloway, lies in that error of omission that Darwin made when he failed to record exactly which of the islands his various finch specimens were taken from. He really needed to know, and he tried to recall details that he should have written down at the time of collection. Darwin could hardly present the finches as his triumphant evidence for evolution, for common descent, when he had no clear idea where he had gotten them, other than on four islands in the Galápagos. If there was one thing Darwin knew, one thing about which he had not the slightest doubt, it was that his words in *On the Origin of Species* would stir up a philosophical, religious, and even political firestorm, a conflagration of indignation, all directed at him. He knew the meaning of what he was saying, he knew what common descent really implied, he knew that natural selection, a mechanistic process, could lead many to the conclusion that a Creator was superfluous—even to the belief that Man created God, not the other way around. For this reason he chose to be extremely cautious about mentioning human evolution in *On the Origin of Species,* a caution he fortunately abandoned later in *The Descent of Man.* Evidence in support of Darwin's reluctance includes an essay he composed in 1842 and a longer essay in 1844, each outlining the idea of evolution and natural selection. His wife was instructed to publish the 1844 essay in the event of Darwin's death.

In all likelihood, the reason why Darwin delayed so long in publishing his notions about evolution and natural selection was that he anticipated and perhaps feared the kind of reaction the book would eventually engender. It wasn't until he was essentially forced to publish, when Alfred Russel Wallace came upon the theory of natural selection, that he rushed *On the Origin of Species* into print, even then calling it "an abstract." And, throughout *On the Origin of Species,* he had to be absolutely solid in his own mind about each and every example that he included. He wasn't rock solid on the finches. He had, in fact, blown his analysis of the finches. Thus no mention of finches found its way into *On the Origin of Species.* Perhaps due to his own embarrassment at his oversight, he apparently chose to ignore the finches when formulating his arguments for his notebooks.

But why didn't Darwin's errors trouble Gould? Gould's sharp eyes for birds saw what Darwin missed, namely that the diversity of these little black and brown birds was only skin deep, so to speak. Gould recognized that the cactus finch was not a member of the oriole/blackbird family as Darwin, upon more casual inspection, assumed and that the warbler finch was not a wren or a warbler. It was a finch. Darwin, upon hearing such a revelation, must have really wanted to get back to those remote islands to take a second look. But Gould had another data set available to him. It seems the young captain of the *Beagle*, Robert FitzRoy, along with his servant, Fuller, made finch collections on the Galápagos, and FitzRoy was careful enough to accurately label those specimens with the island from which each was taken. So Gould could use FitzRoy's notes, and his birds, for island-to-island comparisons among specimens. But first he needed access. Darwin and FitzRoy had a complex relationship, once very close and friendly but which deteriorated as the men aged. Fortunately for Darwin, FitzRoy did agree to make the finch skins available, so Gould got the data he needed. There is much irony here because FitzRoy, an unshakable biblical literalist, was devastated later in his life by the realization that he had provided the means for Darwin to derive his heretical theory. He supplied the ship, and he supplied the finches, at least the most important ones.

Another related mystery remains. Why did Darwin say nothing of the giant tortoises in *On the Origin of Species?* He noted the interisland variation among these huge reptiles during his stay on the islands. Apparently Darwin saw the tortoises as he did the finches and mockingbirds, namely that the variation among them represented variation *within species,* which, indeed, it does for the tortoises. Darwin did not at the time regard such variation as particularly significant. He knew varieties within species existed among animals. It was when Gould told him the finches and the mockingbirds were *separate species* that his attention was focused in a way it had never been before. Beyond that, in the case of the tortoises, Darwin believed that they were not native to the archipelago but instead represented a species that had been brought by sailors from elsewhere, probably to serve as a food source. At the time the Galápagos tortoises were placed in the species *Testudo indicus,* the same name given to the giant tortoise found on the island of Aldabra, in the Indian Ocean. Interesting though they may be, Darwin never regarded them as native and thus paid little attention to them other than for amusement. Sulloway noted that Darwin saw only live dome-shelled tortoises. Perhaps if he had seen live saddleback tortoises, uniquely adapted as they are to reach upward with long necks and procure cactus pads from the tree-sized plants of the genus *Opuntia,* he might have taken greater notice. But apparently he and FitzRoy saw only shells from these animals. Again, as with the finches, it was not until Darwin was back in England that the herpetologist Thomas Bell informed him that the Galápagos giant tortoise was indeed native to the archipelago and was a different species entirely from the one found on Aldabra. What Darwin had quite innocently presumed to be a mere variety, imported from elsewhere, was a whole lot more.

Darwin's most satisfying collection turned out to be the hundreds of plant specimens he procured from the Galápagos. These were turned over to Darwin's close friend Joseph Hooker, the first of his peers to whom Darwin explained his theory of natural selection. It was Hooker's analysis, which was completed in 1845, that Darwin chose to cite in detail in *The Voyage of the* Beagle. Darwin included a table illustrating the high endemism among the plants on the four islands from which he made his collection. His enthusiasm is clear:

> Hence we have the truly wonderful fact, that in James Island, of the thirty-eight Galapageian plants, or those found in no other part of the world, thirty are exclusively confined to this one island; and in Albemarle

Island, of the twenty-six aboriginal Galapageian plants, twenty-two are confined to this one island, that is, only four are at present known to grow in the other islands of the archipelago; and so on, as shown in the above table, with the plants from Chatham and Charles Islands.

Darwin continued, waxing enthusiastically about how *Scalesia* species and other endemics are distributed throughout the islands but found nowhere else in the world. He knew, by the time he wrote *The Voyage of the* Beagle, that the tortoise was a Galápagos endemic. He concluded the paragraph by likening the remarkable endemism of the flora to that of the fauna and mentioned the tortoise, the mocking-birds, the iguanas, and, yes, the finches. Sulloway cited a letter that Darwin sent to Hooker in July of 1845 in which he told Hooker of his total delight in learning of Hooker's conclusions regarding the high endemism of the Galápagos flora. Darwin also stated to Hooker how such results did, indeed, support his beliefs about the distribution and endemism of the animals but added that, prior to Hooker's analysis of the plants, he had always been "fearful" about the conclusions regarding the animals. He was fearful, of course, because he had not taken the kind of care in collecting the animals as he had with the plants. But now he could breathe a sigh of relief.

In the case of Darwin on the Galápagos, truth seems stranger than myth. Darwin never laid eyes on the woodpecker finch, the odd finch species that has adapted to use a cactus spine to poke into crevices and extract insect grubs. Darwin was not impressed by the little brown finches while on the islands. He thought them to be mere varieties and did not recognize their real differences. Mocking-thrushes he believed to be mere varieties at the time and tortoises not native to the islands and of no profound interest. But the plants saved him. Still, Darwin was more confident about analyzing the Galápagos examples in such writings as *The Voyage of the* Beagle than in *On the Origin of Species*. His specimens had not passed muster, and he largely avoided using them in his most crucial analytical work.

History shows that Darwin became an evolutionist for one very clear reason: the Galápagos Islands and the collection of plants and animals taken from the archipelago. And his conversion was anything but quick. It was a long and eventful thought process in which a series of oversights and mistakes must have given Darwin much pause. Darwin had to rely heavily on others to correct and clarify his views regarding Galápagos biodiversity. Darwin, while on the *Beagle* voyage, mused that climate was the major engine of biological change, but he eventually realized that in the Galápagos the climate alone was not adequate to account for the organic variation he observed.

Darwin envisaged common ancestry, a genetic chain in time linking all life ever to have been on Earth with all that likely ever would be. It was Darwin who, upon reading the work of Thomas Malthus, mentally performed an experiment that arguably ranks with Einstein's efforts regarding relativity. This experiment told Darwin how changes in the environment can bring about changes in the genetics of a population and can actually remold organisms over time, a process whereby some individuals survive because they happen to be genetically best adapted to the changing world around them. And they are the ones to reproduce. And new and marvelous forms can result from this ongoing process. Common descent and natural selection, the effect and the cause that account for the grand patterns evident in the history of life, were Darwin's to claim as his own.

So when you visit the Galápagos Islands and the Charles Darwin Research Station, keep in mind the words of Frank J. Sulloway: "Darwin, through his superior abilities as a thinker and a theroetician, made the Galápagos; and, in doing so, he elevated these islands to the legendary status they have today."

About that, there is no myth.

Selected References

Barlow, N. (and C. Darwin). 1958. *The Autobiography of Charles Darwin*. New York: W. W. Norton and Company.

Barrett, P. H., Gautrey, P. J., Herbert, S., Kohn, D., and S. Smith. 1987. *Charles Darwin's Notebooks, 1836–1844*. Ithaca, N.Y.: Cornell University Press.

Browne, J. 1995. *Charles Darwin: Voyaging*. New York: Alfred A. Knopf.

Darwin, C. 1859. *On the Origin of Species by Means of Natural Selection, or the Preservation of Favoured Races in the Struggle for Life*. London: John Murray. [A facsimile edition is available from Harvard University Press, Cambridge, with an introduction by Ernst Mayr (1964).]

———. [1906] 1959. *The Voyage of the* Beagle. Reprint, London: J. M. Dent and Sons.

Desmond, A. and J. Moore. 1991. *Darwin: The Life and Times of a Tormented Evolutionist*. New York: Warner Books.

Keynes, R. D., ed. 1979. *The* Beagle *Record*. Cambridge, UK: Cambridge University Press.

———, ed. 2000. *Charles Darwin's Zoology Notes & Specimen Lists from H.M.S.* Cambridge, UK: Cambridge University Press.

Lambourne, M. 1987. *John Gould: Bird Man*. Milton Keynes, UK: Osberton Publications.

Sulloway, Frank J. 1982. Darwin and his finches: The evolution of a legend. *Journal of the History of Biology* 15:1–53.

———. 1982. Darwin's conversion: The Beagle voyage and its aftermath. *Journal of the History of Biology* 15:325–96.

———. 1983. The legend of Darwin's finches. *Nature* 303:372.

———. 1984. Darwin and the Galápagos. *Biological Journal of the Linnean Society* 21:29–59.

———. 1985. Darwin's early intellectual development: An overview of the Beagle voyage (1831–1836). In *The Darwinian Heritage,* edited by David Kohn. Princeton: Princeton University Press.

5

At Least Two Thousand Craters

The rugged, barren landscape so vividly described in numerous first-impression accounts of the Galápagos is attributable to the archipelago's recent volcanic ancestry. These islands are totally volcanic, and they certainly look it. The darkened rubble that has appeared so forbidding to visitors over the centuries defines the terrain from beach to highlands. Indeed, the Galápagos are in large part a land of blackened beaches and desolate lava fields. The lava itself is hardened magma, igneous rocks left haphazardly behind when the incandescent lava from which they were born finally cooled. To know the geology of the Galápagos, you must know something about volcanoes, how they come to be, and what becomes of them. But volcanoes are only part of a much more intriguing story, one encompassing the geology of the entire planet Earth and one that helped give Darwin his initial insights into the workings of nature.

Earth, along with its sister planets in the solar system, is estimated to be approximately 4.6 billion years old, and the sun is about 5 billion years old. The universe is considerably older. The orbiting Hubble Space Telescope has enabled astronomers to peer back into time at far distant galaxies whose light has traveled through space for perhaps 12 billion years before we have detected it, almost to the big bang itself, when the universe formed. Given such vast measures of time, a million years is short, like a day compared with decades. Life, in the form of bacteria and related organisms, has existed on Earth for about three-quarters of the planet's history, but multicellular life, the complex biodiversity expressed in the forms of plants, fungi, protozoans, and animals, did not become abundant until about 600 million years ago. So for only about 13 percent of Earth's history have there been such creatures as segmented worms, clams, and crabs. Vertebrates are even younger. But in that brief time period Earth has seen millions of species of plants and animals come and go. Whole groups of diverse creatures known by such bizarre names as *trilobites, eurypterids, ammonites, ornithischians, pelycosaurs,* and *condylarths* now exist only in the fossil record. Evolution is unceasing. Why?

The reason is largely found within Earth itself. Unlike our nearest neighbor, the moon, Earth is not geologically inert. Quite the opposite, in fact. Earth is geologically turbulent, a highly dynamic planet. The only place in the solar system more geologically active than Earth is Io, Jupiter's closest large satellite, which, because of extreme proximity to the huge planet, is pulled by immense tidal forces. The Galileo space probe has sent back dramatic photographs of volcanic plumes from severe eruptions on Io's surface. This, the closest of the four Galilean satellites to Jupiter, is a very restless place.

On Earth, the geologic drama is not so obvious but is at least as significant. Whole continents are constantly being fractionated and repositioned as they ride like bulky granitic passengers atop an array of immense and perpetually restless basaltic plates, themselves generated from the thermodynamic engine of Earth's underlying mantle and core. Earth's geology, like its life forms, constantly changes as time passes. Seas rise and fall, mountain ranges erode, and new mountains ascend. What was once verdant forest can in time be replaced by barren desert. A visit to the Petrified Forest National Park in northeastern Arizona will demonstrate this fact with utter clarity. What is now desert is covered by scattered fossilized remains of conifers that were part of a dense forest some 200 million years ago. Throughout Earth's history, there have been times when walls of ice in the form of glaciers periodically expanded and contracted, covering parts of continents and leaving vast amounts of rock and rubble in their wake as they retreat. Earthquakes shatter what looks like solid ground, and volcanoes unpredictably erupt, regurgitating molten minerals that flow like rivers of muddy fire reminiscent of the mythical land of Dante. But unlike Dante's inferno, the internal fires of our planet are not the stuff of myths. Earth's geology is magnificently restless, and that's the key to understanding the generation and regeneration of Earth's vast biodiversity. Through persistent evolutionary change, life generally keeps pace with geological change. And, of extreme importance to life and its history, geological change is a given.

Some places on Earth are more geologically active than others. Californians know only too well that the San Andreas Fault, a complex area of seismic tension, runs all the way from Los Angeles north to beyond San Francisco and threatens at any moment to convert masses of potential energy into renegade kinetic energy, causing what is nervously referred to as "the big one." From southern South America northward through the entire chain of the Andes Mountains, continuing through the western Mexican cordilleras and farther northward through the Sierra Nevada and Cascade Mountains of California and the Pacific Northwest, all the way through Alaska, these various youthful mountain ranges are known for their occasional but persistent volcanism and earthquakes. Mount Saint Helens, which erupted in 1980, is but a recent example of a long-term pattern.

It was not until the mid–eighteenth century that the notion of an ever-changing geology was well articulated. Prior to then Earth's various landforms and seas had been largely viewed as static and unchanging. But the evidence, in plain sight for any geologist who knew how to look, argued persuasively to the contrary. When James Hutton, a Scottish geologist, studied patterns of sedimentation and stratification in such areas as river deltas and coastlines, it became clear to him that Earth not only has changed from what it was in past eons but continues to change. He viewed the planet's geology as in constant but slow flux, due to persistent and widespread forces such as erosion, floods, and volcanism, and this became a model for geology that soon became known as *uniformitarianism*.

The essence of the uniformitarian paradigm is that major changes occur over long time periods by the continual accumulation of small magnitude, gradual, uniform changes in short time periods. Hutton saw Earth as an old planet, always changing, but "with no vestige of a beginning, no prospect of an end." In other words, Hutton's view was that Earth might well be so old as to be eternal. His Earth was steady in time but steadfastly dynamic, with major landforms coming and going from place to place due entirely to natural forces operating on the planet. And Hutton said more. Because Hutton was convinced of the power of small changes accumulated over long time periods, he believed the overwhelming evidence that Earth was far older than the mere six thousand years calculated in 1650 by Archbishop James Ussher, based on the archbishop's careful study of the Old Testament. Hutton's model made Earth subject not only to slow, inordinate change but also to change occurring uninterruptedly for millions, not thousands, of years. It was a radical model, clearly distinct from the notion of a young Earth having experienced a catastrophic Noachian flood, but well supported by data from many regions.

This model of a changing yet steady state geology was further refined and brilliantly championed throughout the nineteenth century by the British geologist Charles Lyell, and it had a major influence on forming Darwin's views on evolution. As well it should. Lyell's first volume of his monumental *Principles of Geology* (1830) was part of Darwin's essential library on the *Beagle*. Other volumes were shipped to him during the course of the voyage. Lyell refined Hutton's views on the antiquity of Earth by suggesting that Earth was far older than traditionally believed from Judeo-Christian scholarship, but not eternal. Earth had a beginning as, presumably, did life. Though Lyell did not seem to accept Darwin's views about the origin of new beings by evolution, he provided Darwin with a great scientific gift, the gift of time. It was Lyell who showed that Earth was, at minimum, millions of years old, sufficiently old for evolution, as Darwin envisioned it, to occur.

Once a static geology was replaced with a dynamic geology, it followed directly that organisms must be periodically faced with some difficult challenges. There's no two ways about it. Given that organisms seem so clearly, indeed elegantly, adapted to the habitats in which they reside (one of the original linchpins of the creationist view), what happens to them when their habitats change and the ecological rug gets pulled out from under them? Good question, but awkward for those who embraced the idea that while Earth itself might change, species had been individually designed, shaped immutably by a creator for their various and specific roles in nature. Logic dictates that once a habitat changes, its resident plants and animals are no longer as well adapted. If a forest becomes a savanna, what is to become of flying squirrels, for example? Lyell thought that they might simply become extinct and that somehow new species, each well adapted to the changed environment, would be quietly and miraculously created by God to replace those who became victims of geologic, hence ecologic, change. Even the notion of extinction was not generally accepted until well into the nineteenth century. Thomas Jefferson, for example, refused to believe that any species could become extinct. Such calamity would reflect badly on the Creator. Darwin had a better idea.

The most obvious parallel to a changing Earth is a changing biota. If evolution is possible, then it becomes clear how life copes with the reality of uniformitarianism. It adapts by evolving into different life forms. Darwin's and Wallace's mechanism of natural selection provides a means by which populations manage in the face of ever-changing environments. Some, of course, fail. Extinction is common to the point where it can be called normal. But adaptation is normal as well, and as Earth's geology restlessly moves the continents about, expands and contracts seas, and builds and erodes mountains, the biodiversity of the planet changes as well. It evolves. As though on a treadmill of time, life must constantly change in order to stay in place. Life forms change, but life itself persists. It has been said that evolution is the "ultimate existential game." You can never win, but only earn the right to continue playing.

It is thus fitting that the Galápagos Islands, the oft-called laboratory of evolution, are situated in one of the most geologically exciting areas on Earth. The two great concepts, uniformitarian geology and evolutionary biology, make very comfortable bedfellows on these young and geologically turbulent islands.

When you walk on any of the islands, know that the volcanic material beneath your feet itself might have arisen from the depths of the sea, escaping as molten magma, cooling and solidifying when it hit the cold, deep Pacific Ocean water, and gradually but inexorably building to reach the surface in a majestic process of geological creation that probably began sometime between 5 and 9 million years ago. Since that time the volcanism of some of the islands has remained active, and new materials have been added. But islands age, volcanoes cease to be active, and thus some of the products of this creation, the earliest islands of the archipelago, are believed to have already subsided beneath the Pacific Ocean, to be replaced by the younger islands that we now visit. The islands slowly come and go, and they move as well. It's an amazing story.

The crust of Earth experiences constant tension due to an ongoing process, unknown until the late 1950s but now well established, called *plate tectonics*. Earth's crust is divided into about sixteen major basaltic plates that move, driven by strong convection heat currents generated from within the radioactive core of the planet and transmitted up through the mantle, the turbulent zone underlying the thin surface crust. Where plates collide with one another, they plunge back into the underlying hot mantle of Earth, only to be regenerated as new material elsewhere. Should a plate containing a continent collide with one lacking a continent, the one without the continent plunges below the one carrying the continent, a process termed *subduction*. An example occurs along the western coast of South America, where the Nazca Plate, which contains the Galápagos, moves east to collide with the South American Plate, which, as you have probably guessed, bears the continent of that name. The Nazca Plate subducts, diving slowly and ponderously beneath the South American Plate. The stunning effect of this colossal collision is the continuous rise of the Andes Mountains and creation of the ongoing earthquakes and volcanism for which the region is well-known.

The power of the plate tectonics model is how well it explains major patterns of Earth's geology. It is the reason for such a close "fit" between the shapes of eastern South America and western Africa. It accounts for the world's tallest mountain, Mount Everest, now considered to be 29,028 feet tall (8,853.5 m). Everest is the result of India, once an island continent like Australia, drifting northeast and literally colliding with southern Asia, causing the uplift of the Himalayas, Mount Everest among them.

As crust is lost in subductions, new crust is formed in other areas. For example, new crust emanates from the immense midoceanic ridge that extends about 20,000 miles (32,180 km) from the North Atlantic Ocean through the South Atlantic and turns eastward and continues north to eventually terminate as the Great Rift Valley in Africa. New seafloor emerges from the ridge moving away from the ridge on either side, like mirror images. The ridge separates the plates containing North America on the western side of the ridge and Europe on the eastern side; thus North America and Europe are moving a few centimeters farther apart each year. Such is plate tectonics and its most notable consequence, continental drift.

Geologists are confident that plate tectonics is real. Continents really do move; new material does emerge from the rifts. An immense database, ranging from patterns of geomagnetism to the present and fossil distribution of animals and plants, provides strong support for what amounts to a new paradigm in geology, plate tectonics, the twentieth-century version of uniformitarianism. Hutton and Lyell would both be thrilled.

All of the Galápagos Islands are situated on the Nazca Plate, one of the major plates of the Pacific Ocean. However, the islands are near an area where the Nazca Plate intersects with two other plates, the Pacific Plate to the west and the smaller Cocos Plate to the north. The Nazca and Pacific plates are moving slowly apart at a north/south juncture called the East Pacific Rift, where new seafloor is emerging. The Cocos Plate is moving north, also away from the Nazca, at a boundary called the Galápagos Rift, another area of seafloor spreading. While the juncture among the three plates is certainly geologically dynamic, it is not what is directly responsible for the islands themselves. That honor goes to the hot spot, the Galápagos mantle plume.

The islands appear to originate largely from the same geographic location and are essentially all volcanic in origin. Only on Baltra, North Seymour, and the northeastern part of Santa Cruz are there marine sediments that were uplifted to make the islands, but even these sit atop volcanic material. If you could tour the Galápagos by submarine, you would soon see that the islands sit on a massive platform of accumulated volcanic material. Nearby to the east is a shallow area, the Carnegie Ridge, while to the northeast lies the Cocos Ridge. The presence of the volcanic ridges and the islands themselves indicates continuous and substantial geologic activity. The islands are actually the tips of amassed volcanic aggregations.

Geologists date volcanic rock using such techniques as the known rates of radioactive decay of various elements. The ratio of the elements in rocks is used to calculate the age of the rocks. The most common technique is potassium-argon dating. Aging the volcanic rocks of several of the islands shows a very clear east/west pattern, in which the rocks from eastern islands date older than those from the western islands. Fernandina, the farthest west, is thought to be only about 700,000 years old, while Española, well to the east, is considered at least 3,250,000 years old, roughly 4.6 times as old as Fernandina. Apparently, youthful Fernandina is situated near the point where the islands originate, while Española has moved steadily eastward since its emergence. Roca Redonda, west of Isabela and Fernandina, is now a very active volcanic site and is predicted to become the next of the Galápagos Islands.

To make sense of this pattern, geologists have proposed the theory of hot spots, more formally referred to as *mantle plumes*. A hot spot is quite literally that, a point where molten magma continuously breaks through a basaltic plate, in this case the Nazca Plate. It is rather like a leak in the core of Earth, permitting material from the underlying mantle to flow through and break the surface crust. The accumulation of hot spot–generated magma creates the actual islands themselves. What is important to realize is that the hot spot is stationary and the plate moves. Thus the hot spot is venting volcanic material that is then transported by the moving plate itself. It is believed that as plate boundaries cross hot spots, such features as the Carnegie and Cocos ridges are formed. New islands are continuously built up at the hot spot as older islands are carried away on the moving plate.

The hot spot or mantle plume originates at the boundary between the core of the planet and the mantle. It can be visualized as a column of very hot rock, a mere 50–60 miles wide (about 100 km), slowly moving upward at a rate of around 4 inches a year (about 10 cm). The rise of a mantle plume is due to the fact that it is hotter (by somewhere around 360°F, or 200°C) than surrounding rock and is therefore less dense. The plume passes through the asthenosphere, part of the mantle of Earth, and collects in large pools called *magma chambers*. Some of this magma eventually breaks through the lithosphere to the surface. When the mantle plume rises toward the lithosphere, it has the effect of quite literally pushing the lithosphere upward. As magma accumulates, and as the plume presses against the lithosphere, the region rises, accounting for the formation of the Galápagos Platform, where the islands lie.

Mantle plumes are relatively common and are thought to account for the formation of the Hawaiian Islands and the Azores.

East of the Galápagos there are submerged, often flat-topped islands called *seamounts,* which appear to be eroded volcanic islands and are now considered quite possibly to be the ancestral Galápagos. A team of oceanographers and geologists headed by D. M. Christie has dated cobbles from these seamounts at ages ranging from 5 to 9 million years, older than any of the present Galápagos Islands. As Christie's team pointed out, if the original Galápagos Islands formed as much as 9 million years ago, there has been more time for colonization and subsequent evolution than once believed. There may well have been tortoises on the islands for longer than any of the present day islands have been in existence! Interestingly, my good friend Peter Alden, who is no geologist but has led many tour groups to the Galápagos, suggested this very point during a shipboard lecture in 1987. Given regular, albeit accidental, dispersal, new islands forming over the hot spot will be colonized even as old islands erode and disappear beneath the sea from whence they came—an example of the relationship between uniformitarianism and biological evolution.

And, in true uniformitarian style, the process of geological upheaval continues. Darwin was most eager to see the Galápagos because he wanted a close look at real volcanic islands. Few, if any, places on the planet could have fulfilled his wish with more clarity. The Galápagos are a superb primer in volcanic geology.

Each of the major islands contains one or more calderas, though the calderas are of differing ages. Isabela has five. Calderas result from the particular kind of volcano that is born of the mantle plume beneath the forming islands. Galápagos magma loses much of its gas when it first meets the sea. The result is that eruptions are mostly lava flows, not huge explosions such as occurred, for example, on Krakatau Island in

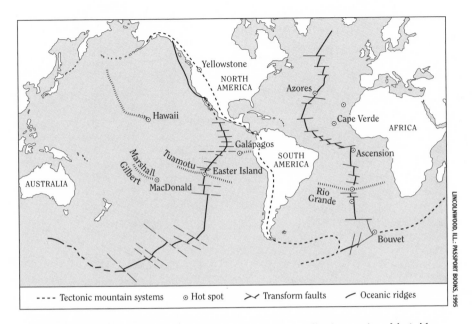

The location of the major oceanic ridge systems where seafloor spreading is occurring. *Adapted from P. Constant,* Galapagos: Darwin's Islands of Evolution

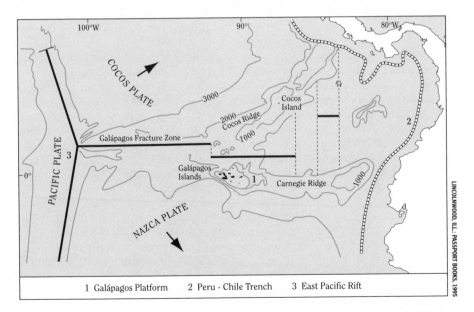

1 Galápagos Platform   2 Peru – Chile Trench   3 East Pacific Rift

The tectonic plates and ridge locations in the vicinity of the Galápagos Islands. Note that the islands originate from a hot spot near the border between the Nazca Plate and the Cocos Plate and close to the Carnegie Ridge. *Adapted from P. Constant,* Galapagos: Darwin's Islands of Evolution

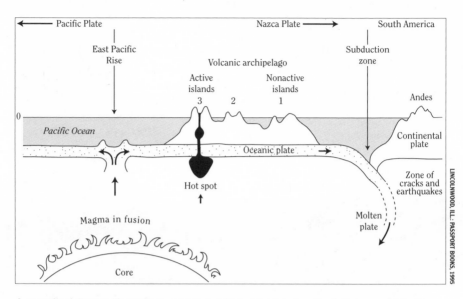

A generalized diagram illustrating how a hot spot forms islands such as the Galápagos and how those islands are moved as part of a tectonic plate. Note that new seafloor emerges from the East Pacific Rise, where magma moves to the surface. The Nazca Plate, on which the Galápagos occur, subducts beneath the South American plate and results in the uplift of the Andes Mountains as well as the frequent occurrence of earthquakes and volcanism. *Adapted from P. Constant,* Galapagos: Darwin's Islands of Evolution

the late nineteenth century. E. O. Wilson, in his book *The Diversity of Life,* provided a concise discussion of the colonization of Krakatau following the eruption.

The typical Galápagos volcano is a shield volcano, so named for its resemblance to a shield with its convex side upward. Shield volcanoes have gentle slopes, not the abrupt conical shape characteristic of typical stratovolcanoes such as Mount Rainier, for example. The Hawaiian Islands, which are also believed to have formed from mantle plume/hot spot generation, are also characterized by shield volcanoes. However, by far the largest of the known shield volcanoes is not on Earth but on Mars! Olympus Mons rises fully 16.4 miles (26.4 km) above Martian "sea level," which makes this immense volcanic mountain nearly three times taller than Hawaii's Mauna Loa, which rises 5.7 miles (9.2 km) above the seafloor. The width of the Olympus Mons crater is an amazing 55 miles (88.5 km). Why is Olympus Mons so immense? The hypothesis is that Mars consists of a single, unmoving tectonic plate containing the hot spot from which Olympus Mons formed. Thus the magma generated through the hot spot continues to accumulate in the same location, forming an ever-larger shield volcano. Perhaps some day in the not too distant future, geologists whose fascination for their subject traces back to Hutton and Lyell will explore Olympus Mons.

The term *caldera,* from the Spanish for "cauldron," is well chosen. Darwin remarked that, to him, Galápagos volcanoes resembled huge pitch pots, the lava having spilled over the sides, staining the gigantic calderas from which it came. Inside the caldera is a broad shelf, the actual roof of the magma chamber, which sits, quite literally, directly over the area of possible future eruption. The caldera walls surround the shelf. Perhaps needless to say, life within a caldera can be subject to periodic upheaval or, more accurately, downheaval. Fractures created by volcanic pressure along the crater walls can undermine the stability of the shelf to the point where it literally drops. Such a dramatic event occurred on Fernandina in 1968 when the crater floor of Volcán La Cumbre, which was 2,624 feet deep (about 800 m), all too quickly descended an additional 1,148 feet (350 m) after an eruptive explosion that launched volcanic ash about 15.5 miles high (25 km) and spread the ashy particles over an area of several hundred square kilometers. That's noticeable. More of the caldera wall collapsed in 1988, effectively widening the caldera.

Vulcán La Cumbra, being on Fernandina, is the youngest, one of the largest, and still one of the most active volcanoes on the archipelago. William Beebe quoted a lengthy account by Benjamin Morrell, who, in February 1825, was witness to one of Fernandina's more dramatic hiccups. Some highlights from Morrell's account:

> The heavens appeared to be one blaze of fire, intermingled with millions of falling stars and meteors; while the flames shot upward from the peak of Narborough to the height of at least two thousand feet in the air.
>
> . . . A river of melted lava was now seen rushing down the side of the mountain, pursuing a serpentine course to the sea, a distance of about three miles from the blazing orifice of the volcano.
>
> . . . the flaming river (of molten lava) could not descend with sufficient rapidity to prevent its overflowing its banks in certain places, and forming new rivers, which branched out in almost every direction, each rushing downward as if eager to cool its temperament in the deep caverns of the neighboring ocean.

Morrell went on to describe how the temperature increased to the point where the air temperature reached 123°F (50.6°C) and the ocean temperature was 105°F (40.6°C). However, when his ship, the *Tartar*, closely approached the lava that was flowing into the sea, the water temperature reached an amazing 150°F (65.6°C). Morrell noted that the heat was sufficiently intense that some of his men collapsed, and he commented on how grateful he was that their little boat was not becalmed in the burning sea.

Most Galápagos visitors have no such thrills in store for them. Instead they must content themselves to walk the desolate but starkly beautiful lava fields, gaze into the quiescent calderas, sift the volcanic sands through their fingers, and contemplate the processes by which these things came to be.

But eruptions do occasionally occur. In September of 1998 Cerro Azul, a 5,541-foot (1,690 m) volcano on Isabela, began erupting for the ninth time in historical record. Recent eruptions, within historical times, are also recorded for other volcanoes on Isabela (Volcán Wolf, Volcán Alcedo, Sierra Negra, Volcán Darwin, and Volcán Ecuador), as well as for volcanoes on Floreana, Pinta, Marchena, and Santiago. It is clear that over the last ten thousand years there have been eruptions on Genovesa, San Cristóbal, and Santa Cruz as well.

The shield volcanoes seen on Fernandina and Isabela are the widest on the islands. Volcán Alcedo on Isabela is close to 5 miles wide (8 km). Alcedo rises to a height of 3,650 feet (1,100 m). Volcán Wolf, also on Isabela, is the highest volcano on the islands, reaching a height of 5,600 feet (1,700 m). Smaller volcanoes are found on the eastern islands. Some, like that on Genovesa, are partly submerged. Others, like the extinct volcanoes of Española and Santa Fé, are significantly eroded.

Most Galápagos visitors quickly learn something of the morphology of volcanoes and the lava that spills from within them. Lavas are really molten metal, and they feel like it. There are often hollow, clanking sounds evident as you walk across an old lava flow. It is as though someone took thousands of old steam locomotives, automobiles, and other scrap metal miscellany and melted them all together.

Most lava is made of hardened, blackened magma and becomes basalt, a rock rich in iron and magnesium, though poor in silicon. Basalt makes up much of the seafloor. But not all lava is basaltic. On Volcán Alcedo on Isabela, the lava is rhyolite, a magma that is highly viscous with abundant water, which accounts for why eruptions of rhyolite lava are typically major explosions such as characterized by Mount Saint Helens. These explosions are of such magnitude that dustlike pumice is formed rather than rocky lava. The last such explosion of Alcedo is estimated to have occurred about one hundred thousand years ago. The white rhyolite pumice can still be seen against the blackened background of basalt on the slopes of Alcedo.

One form of lava is aa (pronounced "ah-ah") and another is pahoehoe (pronounced "pa-hoy-hoy"). Both aa and pahoehoe are Hawaiian words that have been widely adopted to describe lavas. Rough and jagged, aa flows (*aa*, in Hawaiian, means "hard on the feet") are random, uneven, often spiny assemblages of molten rock that formed as gas bubbled out of the lava as it was flowing and cooling. Most of the lava

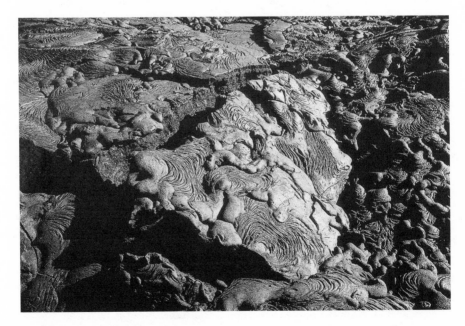

Aa lava, Santiago (James)

Pahoehoe lava, Santiago (James)

on Fernandina is aa. Pahoehoe lava is ropelike, having formed from a rapidly moving, highly fluid lava flow. When you see films that depict lava "rivers," you are watching pahoehoe in the making. The temperature of such a flow can reach 2,000°F (1,000°C) and move up to 30 miles/hour (48 km/hour). When the lava cools, it remains in patterns that resemble ropes and folds, like metal "dough." Some people think that pahoehoe looks like large masses of intestines. Lava can also form as blocks, large chunks scattered about the landscape.

Tephras are often simply called "volcanic projectiles," the collective ash, pumice, and hardened magma thrown from the volcano. Darwin noted what he called *scoriae,* shapeless magma rocks that litter the landscape following a volcanic eruption. Scoriae may be of various sizes, from boulders to fine volcanic sands that are found on beaches such as at Santiago, for example.

Lava tubes are common on the islands and consist of a solidified crust that once was part of a streamlike flow of lava. Tubes form merely because the outer part of the lava stream cools more rapidly than the interior; thus the outer part solidifies first, while the hotter inner part of the flow continues. Once the flow is over, the tube remains.

Spatter cones are small cones, usually with steep sides, that occur in the vicinity of a volcanic vent. If you look carefully at a spatter cone, you can understand its origin. It looks like a petrified lava fountain, and that is pretty much what it is. Small openings permit lava to spout upward, forming the cones. Spatter cones look kind of like little volcano wannabes. Superb spatter cones can be seen from the trail at Bartolomé.

Tuff cones form similarly to spatter cones but are found along the shoreline because they have their origins within the shallow ocean itself.

Darwin noted that not only were Galápagos volcanoes impressive in size but "Their flanks are studded by innumerable smaller orifices." He was likely referring to fumaroles, vents along the slope of a volcano or within the crater where gas escapes. Most of the gas escaping is water vapor and carbon dioxide, but other gases, most particularly sulfur and ammonia, can escape as well. There is a well-known fumarole on the western side of Volcán Alcedo.

Escaping gases emphasize one further point, one with a certain irony associated with it. The Galápagos Rift is located at a depth of about 8,500 feet (2,600 m) below the sea, approximately 215 miles (350 km) northeast of the archipelago. In 1977 humans explored the rift with a submersible vessel and made an astonishing discovery. Life not only existed at this depth but was thriving. Colonies of immense tubeworms and huge mussels clustered together in a tableau that resembled a science fiction writer's creation of life in some distant galaxy. In the utter darkness of the ocean depths, life abounded.

What made the discovery most remarkable was not merely that life was present. A long time ago oceanographers and marine biologists learned that life lived at the greatest depths of the oceans. What was notable about the Galápagos Rift is that it is an area that the hot spot vents heat, an area called a hydrothermal vent. It is an area where oxygen is scarce, pressures immense, light nonexistent, and, most importantly, the vent temperatures can reach 650°F (350°C). Though oxygen is low, the hydrogen sulfide concentration is very high. The combination of heat and acrid hydrogen sulfide simulate hell rather well. But this is no hell. Far from it. Carbon dioxide is also present, and the combination of hydrogen sulfide and carbon dioxide suffice, along with the immense heat, to support a thriving bacterial community, a community of the most thermophilic (heat-loving) organisms known to exist.

The bacteria are termed *chemoautotrophic,* meaning that they are able biochemically to take hydrogen and energy from hydrogen sulfide and combine them with carbon dioxide to make energy-rich organic chemicals. They do in the darkness what green plants do in the light. They use the heat of Earth itself to help catalyze the biochemical reactions essential for life. Mats of bacteria grow densely along the vents where temperatures exceed the boiling point of water. These bacteria are in part the source of the term *extremeophiles,* organisms capable of thriving in what to most organisms would be lethal conditions. The bacterial community is the food base for the tubeworms, mussels, and other multicellular animals. Some of the bacteria associate intimately with the animals, which house the bacteria in specialized cells. The host animal actually uses some of the energy captured by the bacteria for itself.

Some biologists have speculated that life on Earth may have originally evolved from reactions catalyzed by extreme heat such as is found today only in hydrothermal vents but was widely present 3 to 4 billion years ago, when the planet was young and still very hot. The Galápagos Islands and the hot spot that made them may prove more fundamental to understanding the evolution of life than Charles Darwin would have ever imagined, even in his wildest dreams.

Selected References

Bailey, K. 1976. Potassium-argon ages from the Galápagos Islands. *Science* 192:465–66.

Beebe, W. [1924] 1988. *Galápagos: World's End*. Reprint, New York: Dover Publications.

Carson, H. L. 1992. The Galápagos that were. *Nature* 355:2202–203.

Christie, D. M., Duncan, R. A., McBirney, A. R., Richards, M. A., White, W. M., Harpp, K. S., and C. G. Fox. 1992. Drowned islands downstream from the Galápagos hotspot imply extended speciation times. *Nature* 355:246–48.

Constant, P. 1995. *Galapagos: Darwin's Islands of Evolution*. Lincolnwood, Ill.: Passport Books.

Grassle, J. F. 1985. Hydrothermal vent animals: Distribution and biology. *Science* 229:713–17.

Hall, M. L. 1983. Origin of Española Island and the age of terrestrial life on the Galápagos Islands. *Science* 221:545–47.

Lamb, S. and D. Sington. 1998. *Earth Story: The Shaping of Our World*. Princeton, N.J.: Princeton University Press.

Simkin, T. and K. A. Howard. 1970. Caldera collapse in the Galápagos Islands, 1968. *Science* 169:429–37.

Tunnicliffe, V. 1992. Hydrothermal-vent communities of the deep sea. *American Scientist* 80:336–49.

6

These Great Monsters

"These great monsters" is what Darwin called them. But I disagree. Ignore the huge, imposing shells for the moment, pass over the bulky, scale-covered, elephantine legs, and just look at their heads. Their gentle, unassuming faces resemble the face of an uncomprehending ET, the diminutive extraterrestrial, except that these creatures are very terrestrial and, to put it mildly, weigh a good deal more than the adventurous little space alien. Their dry, leathery necks are slender and surprisingly long, their unprepossessing but rugged heads decidedly blockish, squared off, and adorned with liquidy dark eyes and nostrils like twin punctures drilled into a hardened, blunted snout. Toothless, their wide fleshy mouths are bordered by a sharp, horny beak that efficiently shears off twigs, leaves, and cactus pads. Rigid epidermal scales of various shapes and sizes, draped by deep, wrinkled skin folds, add a final touch of homely facial adornment, to my eyes bringing character and dignity to an unmistakably reptilian countenance.

A dignity that deserves respect. For it is these cumbersome creatures whose oddly shaped upper shells provided the appellation for the archipelago. And it is not totally out of the question that a few geriatric members of their lot may have been mere hatchlings when Charles Darwin visited the islands in 1835. I once saw a cartoon of two Galápagos tortoises where one says to the other, "I knew Darwin. Nice guy." In any case, these are the Galápagos tortoises.

The name *galápago* is not the Spanish word for tortoise (*tortuga* is) but refers instead to a saddle. Some of the races of Galápagos tortoises have carapaces (the upper part of the shells) shaped much like a traditional Spanish saddle, a resemblance that obviously impressed early visitors to the islands. In the nomenclature of science, the Galápagos tortoise now is accorded the formal name of *Geochelone nigra* (formerly *elephantopus*), though it was formerly placed in the genus *Testudo*. Most researchers recognize fourteen distinct subspecies or races of Galápagos tortoise, and each of these has its own defining name added to the scientific name. For example, on southern Isabela, you can find *G. nigra vincina* in dry habitats typically at low elevation and *G. nigra guentheri* in more wet habitats at higher elevations.

Face of a dome-shelled giant tortoise at the Charles Darwin Research Station

Probably all visitors to the islands are eager to see the giant tortoises. After all, the lumbering reptiles are the real icons of the archipelago. Truth be told, however, very few tourists actually see tortoises in the wild when they visit the Galápagos. Though tortoise populations are present on nine and perhaps even ten of the islands, there are only two sites open to tourists where free-ranging tortoises can be seen: Volcán Alcedo on Isabela and the tortoise preserve in the highlands of Santa Cruz. The former site requires a rigorous and time-consuming overnight hike (not generally included on the typical itinerary), and the latter site, though more easily accessible, is seasonal and tortoise presence often unpredictable. For a good part of the year the Santa Cruz tortoises are not conveniently located and thus not "seeable." So most tourists must content themselves with becoming acquainted with the Galápagos giant tortoises at the Charles Darwin Research Station on Santa Cruz, where hatchlings, segregated by island of origin, are carefully reared for future release and where adults of several races are kept in spacious outdoor pens, essentially for display. These animals are the ambassadors of their species, and even in such an obvious zoolike setting, they are impressive and wondrous.

Given the indisputable zoological uniqueness of the Galápagos archipelago, it would be logical to assume that behemoth tortoises would occur nowhere else on the planet but there. Reasonable yes, but only partially correct. The species *G. nigra* is, indeed, restricted to the Galápagos, but a very similar and equally large species, *G. gigantean,* can be found on the island of Aldabra, located between the Seychelles and Comoros islands, north of Madagascar, in the Indian Ocean, many thousands of miles from the Galápagos.

Darwin, all youthful 6 feet of him, hopped atop a giant tortoise and rode briefly on its shelled back. Other explorers to the islands have written that even two persons can be accommodated on the huge rounded shell and the animal can still proceed essentially normally. In other words, they're big and they're strong. How big? Male tortoises are substantially larger than females and weigh as much as 600 pounds (270 kg). Females usually weigh no more than about 110 pounds (50 kg). The shell normally measures just over a yard (a meter) in length, sometimes a bit more. Compared with extant reptiles such as many crocodilian species, Komodo dragon lizards, leatherback sea turtles, the anaconda snake, and some of the pythons, the large size of the tortoises is perhaps not so remarkable. But compared with all other extant land tortoises, only those on the Galápagos and Aldabra approach such bulk.

As mentioned previously, tortoises can be found on nine and possibly ten of the islands today. At one time they were even more broadly distributed throughout the archipelago with populations, now extinct, on Fernandina, Santa Fé, and Floreana. My colleague Edward Burtt Jr. has found tortoise bones in a lava tube on Floreana. In addition, the population on Pinta, one of the northernmost islands of the archipelago, is, for all intents and purposes, extinct. Only one male remains and, as yet, no female from the island has been located. This male is Lonesome George, who now resides at the Charles Darwin Research Station. Of the larger islands, only Genovesa and Marchena, both northern and relatively distant from the main islands, failed to be colonized by the enterprising tortoises. The Charles Darwin Research Station estimates that the total number of tortoises inhabiting the islands today is about fifteen thousand, a very far cry from their former abundance. Population sizes vary considerably among islands. The largest total number of tortoises is found, not surprisingly, on the largest island, Isabela, where around 6,600 live, scattered among the five craters. There are also populations on Santiago, Santa Cruz, San Cristóbal, Española, and Pinzón. On Isabela, each population is essentially confined to one of the five craters, and the number of animals differs markedly from crater to crater. The largest single population on the islands is found living on Volcán Alcedo, with about four thousand animals. Volcán Wolf supports around one thousand tortoises, Volcán Darwin about five hundred, Sierra Negra about four hundred, and Cerro Azul about seven hundred.

One of the stories of Darwinian Galápagos lore is that Darwin was deeply impressed by a comment made by Vice Governor Nicholas Lawson to the effect that if Lawson was shown a shell of any giant tortoise, he could tell from which island it was collected. In other words, the populations were markedly distinct, recognizable from island to island. As the story goes, Darwin, once he pondered this information, wondered why any creator would take such pains to sculpt slightly different tortoises from one island to the next. What would be the point in that? Lawson's claim was largely accurate, and it eventually did contribute to stimulating Darwin's post-Galápagos thoughts about evolution. As Darwin wrote in one of his notebooks, "According to this view animals, on separate islands, ought to become different if kept long enough—apart, with slightly differen [sic] circumstances.—Now Galápagos Tortoises, Mocking birds; Falkland Fox—Chiloe fox,—Inglish & Irish Hare." This glimmering into how genetics will diverge among geographically isolated populations made it into *On the Origin of Species,* but the example of the tortoises was not cited.

There actually are sufficient differences in shell characteristics among tortoise populations that an experienced observer can generally identify to which island an animal belongs. But recall that Darwin considered the tortoises to be mere varieties (which, in fact, they are), and he did not realize *G. nigra* was a Galápagos endemic but thought at the time it was the same species, then named *Testudo indicus,* already described from Aldabra. So the comment by the vice-governor may not have been more than a curiosity to Charles.

Many, if not most, of the characteristics that separate the various tortoise populations are genetically based, inheritable from generation to generation. That being the case, it is appropriate to refer to the various island populations as *subspecies* or *races,* rather than merely *populations.* With the exception of Isabela, each of the islands on which there are or were tortoises harbored a distinct race. In the case of Isabela, there are five distinct races, each living on or in a different crater.

Though there are eleven extant and three extinct subspecies, it is easily possible to recognize but two extremes in shell shape. One or the other of these extremes is expressed, to varying degrees, in each of the subspecies. The upper shell, or carapace, may be very rounded, looking like a small land tortoise carapace, only much enlarged. This type of shell is referred to as *domed.* It's shaped a lot like a very large metal army field helmet. Or the carapace may be sharply raised at its anterior (front) end, curved upward in a large distinctive arch-shaped frontal notch, resembling an inverted U. This kind of shell is the one that resembles the Spanish saddle *(galápago)* and is thus referred to as *saddleback.*

There are some important generalizations that distinguish between domed and saddleback tortoises. Domed tortoises, regardless of subspecies, are generally larger in body size than saddlebacks. And domed tortoises tend to be found at higher, cooler, moister elevations, as well as along caldera rims and within the craters. Saddlebacks mainly inhabit lowland arid habitats, the shrub and palo santo zone, where tree-sized cacti are often abundant. Thus islands such as Española, in which there are no highlands, support a saddleback subspecies, whereas islands such as Santa Cruz support a domed shell subspecies. Volcán Alcedo, on Isabela, is also inhabited by a domed shell subspecies so the only place where visitors to the islands can see the saddleback type is in the pens at the Charles Darwin Research Station. The only two free-ranging populations that are permissible for tourists to visit are on Santa Cruz and Alcedo (Isabela) and are domed shell types.

As you might expect (and as I noted earlier), the dome and saddle shapes are extremes in what is in reality a continuum. There are races that are best defined as intermediate between the two types. The domed shape is probably the evolutionarily conservative type. The logic behind this statement is that young tortoises, domed or saddleback, all look quite alike during their first years—they look domed. There is a possible alternative explanation for the domed morph of the juveniles, and that is predator protection. Adult saddlebacks have little to fear, but juveniles are vulnerable. The notch on the shell prevents full protection of the neck and head; thus saddlebacks are, in theory, more at risk. But, in all likelihood, when the saddlebacks first evolved, there were no predators on the islands, so it is doubtful that retention of the domed shape as juveniles is a response to predator pressure. It more likely is the ancient condition. Also, the domed shell type most resembles that of typical mainland tortoises. Founding populations were probably domed, and it was from these that the first saddleback types likely evolved. The most extreme saddlebacks are from Española, Santa Fé, Pinta, and Volcán Wolf (on Isabela).

In addition to shell shape, there are other pertinent distinctions between domed and saddleback tortoises. Recall that domed races are larger in body size than saddleback races. Even more interesting, saddlebacks have proportionately longer legs and necks than dome-shelled animals. The question is why?

When considering questions of anatomical utility in nature, evolutionary biologists first ask if there is any noteworthy correlation between the trait under consideration and the habitat in which the animals live, such that the trait could, in some way, prove to be adaptive to selection pressures exerted on the animals by the habitat. In other words, are saddle-type shells and longer extremities advantageous to tortoise survival and successful reproduction in the lowlands? Do domed shells and proportionately stubbier extremities enhance survival value in the highlands? If not, it is possible that the traits evolved merely by chance and have no influence on the probability of survival and subsequent reproduction of the animals. If this is the case, the trait, whatever it may be, is not an adaptation.

Given that saddlebacks occur in low, arid areas where cacti are often abundant and domed animals are residents of highland areas with high moisture levels, there is a clear habitat separation between the two types that suggests a potential for adaptive significance. Living in the lowlands may require a different array of adaptations from living in the highlands.

Might it be possible that any tortoise, selected at random when an infant at the Charles Darwin Research Station, could be placed on any island, and it would become saddleback or domed depending upon if it grew up in the arid lowlands or moist highlands? If this were the case, we would conclude that shell shape, body size, neck length, and leg length are all determined by the particular kind of environment in which the animal grows up. In other words, the traits, though adaptive, are due to phenotypic plasticity rather than determined by the genotype of the animal. A familiar example of phenotypic plasticity is shown when humans produce higher concentrations of red blood cells when living at high elevations in rarefied air. But the characteristics that distinguish domed from saddleback tortoises are largely genetic, with little environmental influence. Regardless of environment, saddleback tortoise hatchlings only grow into saddleback tortoises. The same holds with the domed races. That's why the tortoise populations are considered racially distinct. Therefore, the body morph, saddleback or domed, is, in all likelihood, a genetic response to differing selection pressures between lowland and highland ecosystems: Body morph has evolved by natural selection and is adaptive in the Darwinian sense. But what might these alleged selection pressures be?

Before focusing on differences, we would do well to begin by considering one obvious and very important similarity between saddleback and domed races. Both are large, indeed very large, well deserving the popular name "giant tortoise." No tortoise species now living on mainland South America approaches the Galápagos species in body size. However, extinct species found on several continents, including South America, are nearly or equally as large. So the ancestor of the Galápagos tortoises may have already been a relatively giant tortoise. Or not. We simply don't know.

If the initial colonizer (or colonizers) was small, it is easy to make an evolutionary argument for rapid size increase with each generation. And, in fact, it is most likely that the ancestral tortoises were anything but giants. Recent work on tortoise DNA has pointed to a small South American species as the likely ancestor of the Galápagos giants. It is reasonable to suppose that a smaller tortoise might have a better chance of traveling by rafting to the islands than a heavier animal. Beyond that, gigantism is a frequent characteristic of island species. Without other species as competitors and with possibly no predators, the tortoises would likely only compete among themselves for food and for mates (the latter of which, initially at least, could have been in very short supply), or both. In such a situation large body size would prove adaptive, and any genetic tendency toward size increase would have high fitness and be disproportionately inherited. Given the impressive size difference between males and females, it is likely that male competition for access to mates has been a major factor in tortoise evolution on the islands. Suppose that throughout tortoise history on the Galápagos, big male tortoises supplanted smaller male tortoises and thus size was selected; that is, it is adaptive. Even if the initial colonizing tortoises were large, the previous argument would still apply. So why is the species not larger still, perhaps twice its present size?

Counter–selection pressures eventually act to halt trends such as body size increase. In the case of the Galápagos tortoises, extremely large animals may have difficulty consuming sufficient food. They may be less able to move long distances, which they must do to find water. They may be unable to mate properly, being too heavy, too awkward. They may be more subject to injury. There is no evidence that the tortoise populations anywhere on the islands are now increasing or decreasing in body size. So the difference in body size, though slight, between domed and saddleback races must be selectively maintained by some factor in their environments.

Domed races live in the highlands and calderas, where it is considerably cooler and wetter than in the lowlands. Tortoises, being reptiles, are ectothermic, unable to metabolically generate body heat. Field experiments have shown that, even as air temperature fluctuates, adult Galápagos tortoises are capable of maintaining a relatively stable body temperature by behavior alone, essentially by moving into sunlight to warm up and ambling into shade or water to cool down. In a cool environment, where heat loss is a major concern, it is generally best to be large. It is also beneficial to have short extremities in such a situation.

The reason why bulk is helpful in cool environments becomes clear by noting the physical relationship between volume and surface area. Large objects, which have proportionately more volume, will lose heat less rapidly than small objects of similar proportions that, as you probably have guessed or already know, have proportionately more surface area. For example, small ice cubes melt more rapidly than large ice cubes (even though both are, indeed, cubes). Surface area is where heat is lost, and thus animals with much surface area will lose heat more rapidly than those with little surface area, just as finely chopped ice melts more rapidly than cubed ice. The larger body size and proportionately smaller neck and limb lengths of domed races of tortoises result in maximizing volume and minimizing surface area and thus may be a response, at least in part, to selection pressures imposed by cool temperatures in the highlands, an area drenched by the garúa mist for much of the year.

And, by the same token, the long necks and legs of saddlebacks could help facilitate heat loss in the sunny, sweltering lowlands. The overall smaller body size of saddlebacks would retain less heat than the larger bodies of the domed animals, also a useful adaptation in a hot climate.

So we have the beginnings of an explanation for the difference in body morphs between domed and saddleback tortoises. But the full explanation is more complex than simple adaptation to temperature.

Domed-shelled animals inhabit ecosystems in which there is often dense undergrowth of vegetation—the transition, scalesia, and miconia zones. The domed shells, smoothly rounded as they are, may prove adaptive as the tortoises move, tanklike, through the dense plant cover, which, of course, is also the animals' food source. On the other hand, saddle-type shells, with the large forward notch, can actually become snagged in low vegetation, impeding the movement of the tortoise. Saddle shells are not adaptive in low, dense vegetation.

In addition, one of the most important food plants for the saddlebacks is tree sized: the cacti of the genus *Opuntia*. These animals, which inhabit more open ecosystems than the domed races, often must reach up to attain access to the cactus pads located some distance off the ground in the tall cactus plants. A long neck and long legs are obviously advantageous in such a situation. The arching notch in the front of the carapace nicely accommodates the neck, permitting even greater upward reach. Thus the saddle shell shape could reflect adaptation to a particularly important food source, one that is never found at higher elevations where the domed animals occur. Some evolutionists have speculated that tortoise herbivory was a major selection pressure favoring evolution of the tree-sized growth pattern of the *Opuntia* species of cacti. In other words, the cacti grew tall to escape the hungry mouths of the tortoises (or, to say it correctly in the jargon of evolutionary biology, the tallest cacti were the most fit, tallness had a genetic basis, so taller individuals survived to reproduce more offspring than shorter plants, and so tallness evolved to become a general characteristic of the species). Some circumstantial support for this evolutionary scenario is evident on islands such as tiny Champion, off of Floreana. Tortoises have never colonized Champion, and the prickly pear cacti grow low and spreading, not treelike. But on islands with both cacti and tortoises, the tortoises, in response to the changing growth form of the *Opuntia* species of cacti, evolved longer necks, longer legs, and a notched shell, in an evolutionary arms race driven by access to cacti. These traits were not problematic in a hot, arid environment where heat loss was also facilitated by longer extremities.

The most fundamental process in all of natural selection, as it applies to Galápagos tortoises, is how Galápagos tortoises manage to make more Galápagos tortoises. *How do they do it?* is a question that probably passes through the minds of most visitors to the Charles Darwin Research Station when first they gaze upon a group of creatures that look, for all the world, like marginally animated boulders. Like all land animals, fertilization is internal. A male, all five hundred or six hundred pounds of him, must mount a female, insert his relatively modest penis (located, logically enough, at the tail, between the hind legs), and ejaculate tortoise sperm. This takes some doing and consumes a fair amount of time even after the male jockeys himself into position. Once in position, males can spend anywhere from a half hour to several hours copulating. Females seem relatively patient. Neither sex appears to be in

The long neck of the saddleback tortoise adapts it to reaching for leaves.

a hurry. And the act of mounting is far from simple. Males often slide off, sometimes even inadvertently landing upside down, at which point they must right themselves and try again. Be assured that tortoise brains are probably insufficiently complex to permit embarrassment. Male tortoises, apparently frustrated or confused, have been observed attempting to copulate with smaller males as well as occasionally with large rocks. But anatomically challenging though it may be, they do succeed (at least some do). After the consummatory act has been performed, the male wanders off, perhaps to offer his services to another female, and it is left to the newly mated female to do the rest. But what is of most interest to us at this point are the events that occur immediately before a male gets to mount a female. In nature, obtaining access to the opposite sex is often the most important part of the life cycle. So it appears to be with giant tortoises.

As I mentioned earlier, in island populations spared from interspecific competition and significant predator pressure, intraspecific competition can quickly elevate in importance. Darwin correctly noted in *On the Origin of Species* that the most likely competitor for any animal is another of its same species. Obviously both have essentially identical ecological needs. And even more obviously, both have the same reproductive needs. In the case of a male Galápagos tortoise, what is needed is a female Galápagos tortoise. Females thus serve to promote male-male competition. And that's when the trouble begins, at least for the males.

Male tortoises fundamentally engage in battle for females. Some accounts call these largely ritualized contests "mock battles," but that is only partially correct. Tortoises can bite. Because they often appear to be in a state of semisuspended animation when supine in their pens at the Charles Darwin Research Station, it may be difficult for most who observe these animals to believe that they can become aggressive toward one another. But testosterone can induce powerful drives. Breeding season is during the rainy months, from about January to June. At this time male tortoises abandon docility for confrontation. Testosterone captains the reptilian brain and focuses it on a single goal. Galápagos tortoise males are well-known to be agonistic toward one another during breeding season. As with most animal species, much of the aggression is ritualistic, but real fighting in the form of biting occurs regularly. And, when two males mix it up, the most important characteristic in predicting if a given male will prevail is just how far he can raise himself off the ground and stretch his neck, a "vertical extension of neck and forelimbs," as a researcher once put it. In Galápagos tortoise society, tall guys win over short guys.

The entire ritual is, of course, more complex. Often two tortoises are of nearly equal vertical extensions (which, from this point onward, I'll simply call "height"), and thus the engagement is not easily broken off. If height doesn't settle the contest, males then gape at one another, opening their mouths widely, presumably looking as ferocious as a small-headed, toothless reptile can look. They emit a bellowing vocalization, the only sound that ever comes out of a Galápagos tortoise other than a loud hiss when one contracts its head and legs rapidly. They expose the intense colors on their necks (tortoises, like all reptiles, see in color). If none of that works, they attempt to bite each other's forelimbs and hind limbs. The loser may break off at any time, with any of the following signals: lowering of the head and neck, lowering of the body, hissing, turning away, or, best of all, straightforward retreat. In experiments in which smaller saddleback males have engaged larger domed males, the saddlebacks usually win by virtue of their added vertical extension.

The picture that emerges is, I suspect, something like this: The principal selection pressure in the evolution of the saddleback races has been male-male competition for access to females. Because lowland races, living as they do in an arid, hot environment, are usually under selection pressure to lose rather than retain body heat, the elongation of neck, tail, and limbs, as well as smaller overall body size, are complementary to selection for vertical extension. Indeed, smaller size may be advantageous in lowland areas because it requires less energy and water and the animals can make better use of the limited shade available. Agonistic behavior of saddleback males may serve to disperse the animals more widely, thus reducing competition for limited food resources. But highland races are in a more food-rich environment, they are not liberated from selection pressures for heat retention, and mobility would be compromised by a saddle-type shell; thus these tortoises have not followed a similar evolutionary pattern. The height of the tree-sized, arid zone *Opuntia* species of cacti may be related, at least in part, to herbivore pressure exerted by the increasingly taller giant tortoises.

Female tortoises, whether domed or saddle shelled, are always smaller than males. But even their size could be evolutionarily related to male-male competition. If larger males were historically at a selective advantage, their genes for size were passed on to their daughters as well as their sons, and thus female body size undoubtedly increased, though less dramatically, as male size increased.

Female giant tortoises, like most reptiles, lay eggs. Females have mated by the end of the rainy season and are then faced with the task of nest building and egg laying. So anytime from February through May females slowly and methodically move into the arid zone to prepare nests. Obviously, saddleback females have little distance to travel, but domed females must migrate from the highlands to lowlands. Egg laying occurs anytime from June through December. Each female may lay several clutches, and clutch size is variable, from two to twenty eggs. Average clutch size varies somewhat among subspecies. In general, domed subspecies lay larger clutches than saddleback subspecies. Egg size has been described as somewhere between a billiard ball and a tennis ball.

In order to lay her eggs, a female must first prepare a nest. This procedure is done by digging into the soil with the hind legs, making a hole approximately 12 inches (30 cm) deep and 8 inches (20 cm) wide. Females work slowly, and the entire operation may consume anywhere from four to as much as twelve hours, depending upon which island the female resides. Females typically urinate on the ground, softening the substrate. When the eggs are laid, the female will seal the nest with a cover of substrate cemented by urine. The mother then leaves the completed nest with the next tortoise generation sealed safely inside. She will have nothing more to do with her offspring. Eggs must incubate anywhere from four to eight months. Most infant tortoises first see the light of day sometime between November and April. To do so, they must first hatch and then claw their way out of the nest, a method traditional for many reptiles. Some babies fail in this task because their nest is too well sealed to permit them to excavate. But many, and probably most, succeed. Once liberated from the nest, baby tortoises must avoid predators and find food. Bear in mind that all of the members of highland races hatch in the lowlands and therefore must eventually wander to higher elevations.

Human parents are typically daunted by how small their newborns are and how much they must grow before attainment of adulthood. These parents should be glad their babies are not tortoises (and I am confident that they are). A human infant girl born at, say, 6 pounds will probably grow into a woman weighing, say, 135 pounds (61 km). At birth, the human baby is somewhere around 3 and 5 percent of its adult body weight. Puberty will occur around year twelve or thirteen. Galápagos tortoise babies, at hatching, weigh only about $\frac{1}{1,000}$ or less of what they will weigh as adults. And sexual maturity will not be reached until the animal celebrates its twentieth birthday, at the earliest. Some giant tortoises are not sexually mature until about age thirty. Though it requires between two and three decades before tortoises mature, adults do tend to enjoy a lengthy tenure on the planet. It is now estimated by researchers at the Charles Darwin Research Station that a giant tortoise can attain an age of 150 or more (though such an extreme may be relatively rare); hence, my comment at the beginning of the chapter about some contemporary tortoises perhaps having been alive when Darwin visited the Galápagos in September of 1835. Had they hatched that summer, any survivors would be somewhere between 160 and 170 years old. It's a stretch, but it's possible. I'd like to think so. There are few animal species on Earth that normally live longer than human beings, but Galápagos tortoises are one such species.

In recent years it has been learned that in many reptile species' sex is determined not chromosomally, as in birds and mammals, but essentially by the temperature at which the eggs are incubated. This appears to be the case with Galápagos tortoises. Research at the Charles Darwin Research Station has shown that embryos become males if the incubation temperature is less than 83°F (28.5°C). Higher temperatures influence embryos to become female. Both males and females can be produced in a single nest because the actual egg temperatures may vary somewhat based on the position of the egg in the clutch. It is obvious, of course, why female domed tortoises must migrate to the arid lowlands for nest building. Were they to nest in the cool highlands, eggs would take much longer to incubate, many would perish from the cold temperature, and any that hatched would be males. Not a good way to sustain a population.

As with most animal species, mortality rates are highest during the juvenile period. Eggs and hatchlings are much more subject to predation than immature or adult tortoises. Most of the predation on eggs and baby tortoises is practiced by species introduced to the islands by humans. Rats, dogs, and pigs are the principal culprits because they will dig up nests and consume the eggs. Dogs kill and devour baby and juvenile tortoises. In many cases, conservationists at the Charles Darwin Research Station have had to resort to finding nests, removing eggs, and hatching the eggs in the safety of the research station. Tortoises are reintroduced on their island of origin when they have attained sufficient size to avoid predators. Prior to the introduction of the predators mentioned previously, the only real predator of baby tortoises was the Galápagos hawk.

Given what amounts to a historical lack of predator pressure on the Galápagos, it is easy to imagine how the initial tortoise populations must have rapidly burgeoned. Evolutionarily, the world's tortoises, like most animals, are adapted to an environment in which egg and infant predation is a significant possibility. Mainland populations are never free from such selection pressures. The presence of predators is a given. But once they colonized the Galápagos, the forerunners of today's tortoises were essentially free from nest marauders. No one knows when the hawk colonized. It is quite possible that no predators were present when tortoises first arrived and began dispersing among the islands.

In order for a sexually reproducing population to remain stable (no net loss or gain), each member of the pair must reproduce itself. That is, there must be a daughter to replace mom and a son to replace dad. (Obviously it need not be that precise, but it does need to average out that way over the entire population.) So, put another way, over the entire reproductive lifetime of a female tortoise, she need only have two of her offspring reach adulthood and successfully reproduce themselves in order for the population to remain stable. Should more than two of her young survive to reproduce, the tortoise population will grow. Now consider that a female tortoise may conservatively lay five to ten eggs annually for perhaps eighty years or more. Even if she averaged only one annual clutch of, say, three eggs per year, this would amount to over two hundred eggs in a lifetime. More realistically, the figure could easily be four or five times that large. So without any real predators at any life cycle stage, most baby tortoises would hatch and grow to become reproducing adults. On average, far more than two animals per female would be produced. The islands would quickly become very full of giant tortoises, which is exactly what mariners first beheld when the islands were discovered in the mid–sixteenth century.

Tortoises are a type of turtle, and turtles are among the most evolutionarily ancient and venerable of the various reptilian groups. Turtles were around throughout the Mesozoic, when giant dinosaurs were prevalent, and turtles persisted through the great Cretaceous extinction event that occurred about 65 million years ago, ending the so-called Age of Reptiles. Apparently no one thought to tell the turtles. Today there remain 230 turtle and tortoise species that successfully inhabit fresh and salt water, as well as all manner of terrestrial ecosystems, from deserts to forests.

The secret to their lasting success is probably their unique anatomy. Like ancient knights in armor, these animals are well protected within a hard casing both dorsally (their carapaces) and ventrally (their plastrons). Their mobility, such as it is, is greatly confined by their protective covering of fused bone and overlying thick scales. But, when danger threatens, the turtle just pulls in and closes up shop, the head contracted, legs and tail tucked tightly beneath the shell. The turtle, with a patience derived from a long evolution and a short intelligence quotient, merely waits it out. It's a strategy that doesn't always work, but it works satisfactorily enough that the world is still reasonably well endowed with turtles, as it has been for nearly 200 million years. Indeed, there have been turtles far longer than there have been primates, elephants, or birds. But, throughout their long history, turtles have generally been assigned the roles of supporting players in the ecological play that is nature. They have never occupied positions of ecological preeminence—until they reached islands.

It is on the Galápagos and Aldabra that tortoises ascended to the ecological niche of dominant terrestrial herbivore. Not since the dinosaur-dominated Mesozoic, when long-necked sauropods, herds of ornithopods, and bulky ceratopsians dominated, has a reptile species been the principal vertebrate consumer of plants in any ecosystem. But, on islands without mammals, such opportunities still exist, even for giant tortoises.

Though many turtle species are predators, tortoises are mainly plant eaters, and the Galápagos giant tortoises appear to be entirely herbivorous. During my most recent visit to the Charles Darwin Research Station, I watched as a maintenance crew unloaded large stacks of freshly cut plants, the salad fixings for the resident adult tortoises. Apparently the tortoises were watching as well. What looked like a slow motion landslide of boulders began converging on the pen's gate, through which the food would be supplied. Tortoises massed together, necks stretched upward, eyes upon the workers. Mentally slow perhaps, but these creatures at least have learned to recognize their waiters.

Galápagos tortoises are fundamentally generalists in their vegetarian diet. Domed species, residents of upland areas, feed heavily on the lush grasses and various of the broad-leaved herbs and shrubs that grow prolifically from the transition through the miconia zones. They eat the endemic scalesias and guayavita *(Psidium galapageium)*. Many researchers have noted that tortoises have a particular fondness for the poison apple *(Hippomane mancinella)*, which, of course, is palatable to them, as well as the various lichens and the bromeliad *Tillandsia insularis*. Saddleback tortoises, as previously discussed, feed heavily on cactus pads and fallen cactus fruits of the genus *Opuntia*.

Tortoises, from a study done by Charles M. Rick and Robert I. Bowman in 1961, also appear to be an important species in the dispersal of the endemic Galápagos tomato *(Lycopersicon cheesemanii)*. Experiments with the tomato showed that the seeds will remain dormant and not germinate unless the tough seed coat is removed. One way for that to happen is to expose the seeds to acidic conditions, such as can be found in stomachs, for instance. A tortoise is a historic symbol of slowness. In the case of the Galápagos tortoise, it's the inside slowness that counts for tomatoes. It requires from one to three weeks, and quite possibly longer, for a tomato seed to make its tortuous way through a giant tortoise intestinal system, a slow peristalsis to put it mildly. But, when the seed finally emerges at the other end of the tortoise, three events have occurred, all good for the seed. First, it can now germinate and grow into a new tomato plant because the coarse seed coat has been removed, courtesy of the biochemistry of the tortoise's innards. Second, the seed is well away from its parent plant because the tortoise has undoubtedly wandered a considerable distance from where it ingested the tomato fruit. Third, the seed is deposited with abundant nutrients available, courtesy of the tortoise's intestinal system.

Tortoises, therefore, appear to be essential seed dispersers for the Galápagos tomato. The tortoise may have even been instrumental in dispersal of the tomato among the islands, though there is no solid evidence one way or another. Further, the researchers could identify no other animal on the Galápagos that could both remove the seed coat and efficiently disperse the seed, though recent work suggests that the Galápagos mockingbird may also disperse the seeds.

Though Galápagos tortoises are large, their ectothermic metabolism means that they need not take in nearly so much food (per unit body weight) as a similar-sized endothermic mammal. So, all things considered, the average adult tortoise has a pretty easy day. They don't need to graze constantly, and so they don't work all that hard. Their daily behavior has been thoroughly documented, and accounts suggest that they often do not begin feeding until well after sunrise. They will continue feeding, albeit at a leisurely pace, until about an hour or so prior to sunset, when they retire for the night to a quiet pool or protected area within a clump of shrubbery. Tortoises must have water, lots of it, and will wander long distances to obtain it. Their winding, well-blazed trails are easily seen in areas where they are common. Highland races often gather in muddy pools where they spend many hours partially submerged and do basically nothing. Some observers have suggested that such behavior helps reduce annoyance by mosquitoes or parasitism by ticks. For whatever reason, they like pools. A thirsty tortoise will consume a great volume of water and will be able to store it for a protracted period. Darwin, adventurous fellow that he was in his younger years, was moved to comment on the taste of the water contained in a tortoise bladder: He described it as having "a very slightly bitter taste."

Darwin was also moved to comment on the relative speed exhibited by the gargantuan reptiles. He estimated that an adult could actually walk 4 miles (6.4 km) a day, even allowing for a lunch break or two. Darwin's figures for tortoise speed have held up: The big animals indeed can cover respectable distances, particularly when in search of water. Tortoise motion is advantageous because it is the movements of the appendage and neck musculature that alter the chest volume, thus expanding and contracting the animal's lungs, allowing it to breath. Unlike mammals, tortoises lack a muscular diaphragm, and given that their ribs are fused with their shells, they must rely on other muscles with which to breath.

Tortoises in the grassy highlands are often used by colorful scarlet and black vermilion flycatchers as perches. The birds, which feed on insects captured in the air, probably use the tortoises as "beaters" of a sort and snatch up insects disturbed as the bird's perches amble along.

A Darwin's finch (such as a small ground finch, for example) may hop atop a tortoise and begin clambering around on the animal's head, neck, and legs. The finch may be picking off seeds that have become stuck to the tortoise as it perambulates through vegetation. But finches also have been observed removing ticks from the big reptiles. Tortoises have developed a helpful behavioral response to the stimulation of a finch on the neck or legs. A tortoise will elevate itself, its neck stretched, its legs extended to their fullest, a behavior that obviously facilitates avian access to the attached ectoparasites. I watched at the Charles Darwin Research Station as our park service guide took a small branch and gently, even tenderly, stroked a reclining adult tortoise on its neck. The animal gave the impression that it was operated by hydraulics: It slowly but steadily began raising its ample self ever upward on its legs. Its neck, like a sluggish crane, went into motion and began to methodically elevate its head, and within a minute or so, the creature had reached full extension. We found no ticks.

Though the conventional wisdom now is that there is but one species of Galápagos giant tortoise, that was not always believed to be the case. Early researchers accorded species status to various morphologically distinct populations. After all, anyone can see that the Española saddlebacks look markedly different from the Santa Cruz dome-shelled animals. Given that various island populations of the animals have such obvious morphological differences, why not recognize them as separate species rather than subspecies? Excellent question.

Biologists today differ in their opinions as to how to define a species, but the majority still accepts a species concept based on reproductive isolation rather than mere difference in morphology. In other words, a species is a population or group of populations that can freely interbreed and produce fertile offspring within its own ranks but cannot and does not breed with other populations. Marine iguanas don't ever breed with land iguanas. They are separate species. But populations of marine iguanas on Fernandina, for example, are thought to be interfertile with those on Española (though they are morphologically distinct, with one larger and one smaller, one gray-black and one much more reddish), so both populations are placed within the same species. Likewise, biologists believe that tortoise populations on separate islands are not sufficiently genetically distinct to prohibit interbreeding, if they but had the opportunity. Though separate races exist, there is thought to be but one species. Given more time, that status could change, especially if tortoise populations are thoroughly isolated from one another so that there is absolutely no opportunity to exchange genes among races.

There is an obvious weakness in this species concept. Since tortoise subspecies are isolated geographically from one another (a distribution pattern called *allopatric*), how do we know that tortoises on separate islands can successfully interbreed if given the chance? In many cases, we don't. We make an educated guess based on the following clues. Tortoise courtship is not highly variable among races, and males of different races, when confined together, will eagerly display at one another, a strong suggestion that they consider each other to be of the same species. A logical strength of the biological species concept, as it is termed, is that the organisms themselves often show us if they are separate species by their behavior toward one another. More importantly, recent genetic studies have demonstrated very little genetic separation among tortoise races. Biologist James L. Patton found that, on average, there is only about an 8 percent genetic difference among populations, which may seem to be a small amount given the clear morphological differences between saddleback and domed populations. But consider that human beings and chimpanzees differ by less than 2 percent genetically and they look considerably different from each other! As with humans and chimps, morphological distinctions between saddleback and domed tortoises could very well be accounted for by a small number of regulatory genes that affect the pattern of development of the animal.

Because of the strong genetic similarity among races, it is likely that there was but one colonization event of the archipelago itself, probably on San Cristóbal or the more ancient islands that are now submerged seamounts. The fact that Galápagos giant tortoises have little genetic similarity to extant South American mainland tortoises suggests that the colonizing species may have become extinct.

From San Cristóbal, the colonizing tortoise population gradually dispersed among the islands. It is considered likely that the saddleback morphs evolved independently (from the ancestral species) on the various islands where they occur or formerly occurred. Dispersal of tortoises throughout the archipelago is, by any reasonable measure, the result of a series of low-probability events. Tortoises that are placed on a beach are sufficiently aware that they are members of the terrestrial community such that they never plod their ponderous selves into the surf, like shelled Magellans. Quite the contrary, they turn and wisely head for the interior. But if a tortoise, large or small, should, by some unfortunate act of nature, find itself washed into the foaming brine, it can float and even propel itself. That's correct, they appear to be able to swim.

William Beebe commented on the facility exhibited by a tortoise that he placed in the sea to test its skill as a mariner. It passed with distinction. Beebe reported that this animal died a few days after its swim, and he attributed its demise to taking in too much salt water, but such a claim was never actually demonstrated. Its death may be entirely unrelated to being in the sea. It is thus quite within the range of credibility to suggest that the animals have, indeed, spread among the islands by their own abilities to survive in the sea, often as passengers on rafts of vegetation torn from the landscape during severe rains and floods. That's how the great ancestor tortoise (if only one, presumably a pregnant female) may have gotten to San Cristóbal in the first place. As for not readily taking to the ocean, most biogeographers assume that colonizing tortoises were accidentally carried from one island to another, and one researcher has suggested that tortoises might have been transported on pumice rafts that became adrift following volcanic eruptions.

For perhaps several million years giant tortoises prospered, increasing in population, dispersing among islands, enjoying peace and tranquility, protected by distance from threats posed by mammalian predators. We humans changed all that. The history of tortoises and humans has much in common with that of the American bison and humans. Once so abundant that no one could imagine their loss, bison were nearly hunted out of existence. From an original population estimated at somewhere around 60 million, bison were reduced to a few scattered, tiny herds. But, though driven to the brink of extinction, they were saved. Perhaps Galápagos giant tortoises have been saved as well.

No one knows what the original population size of the giant tortoise was, but it certainly was not in the millions. The islands are too small in area to sustain such numbers. But no careful censuses were made, so the former abundance of the tortoises can only be indirectly estimated by an analysis of their persecution and subsequent decline. To do this requires examination of accounts and logbooks of mariners—the pirates, the whalers, the sealers, and the explorers—to get some sense of how many tortoises were taken for food or just outright killed. From those admittedly shaky figures plus various early eyewitness accounts from the islands, the conclusion is that conservatively somewhere around a quarter of a million tortoises, perhaps more, were resident on the archipelago before humans, rats, dogs, pigs, goats, and donkeys cut them down to almost nothing. The fifteen thousand remaining animals, although not a small population by the usual standards of threatened or endangered species, represent at most about 5 or 6 percent of what was once present. Just to put this figure in perspective, I live in the town of Bourne, on Cape Cod, in Massachusetts, where during the peak of tourist season the human population is about forty thousand, or not quite three times the total world population of Galápagos tortoises. The human population of the nearby Boston metropolitan area is about 4 million, or about 267 times the world population of Galápagos tortoises. The world population of humans, just over 6 billion, is roughly 400,000 times as large as the tortoise population today.

Tortoises have a really unlucky combination of characteristics, which almost doomed the poor creatures to a quick extinction. They were abundant in number, each animal contained an ample supply of meat, they could not protect themselves, they were easy and safe to capture, they lived a long time in captivity without food or water, and they tasted good. Mariners simply captured them, hauled them aboard, and kept them alive (with no food or water) in the cargo holds for reputedly up to eight months until the animals were eventually butchered and consumed. Beebe quoted an account in which tortoise meat is compared in flavor with "the finest veal." In addition, tortoises yield a high-quality oil and were described as "the most delicious food we had ever tasted." Because the oil is of such high quality, tortoises were also hunted just for their oil. Reports vary, but it was apparently not uncommon for a single ship's company to collect up to three hundred live tortoises. And not all tortoises were taken aboard ships. Many were killed on the islands, often for only a small portion of meat, the remainder left to rot or be consumed by scavenging dogs. Packs of wild dogs, even today, remain a threat to tortoises. Dogs will not hesitate to attack even adult tortoises and tear the flesh from their legs.

At the rate of predation suffered by tortoises in the eighteenth and nineteenth centuries, humans alone would eventually have annihilated the creatures. But with humans came other predators as well as competitors. Pigs root up nests and consume eggs and young. Dogs and rats do as well, and dogs kill juveniles and adults alike. Goats and donkeys compete with tortoises for forage. Hoofed animals of any sort, including cattle, inadvertently step on nests, crushing eggs. If any species of animal was ever entitled to say "there goes the neighborhood," it was the Galápagos giant tortoise the day those first sails appeared on the horizon.

Even as tortoise populations were crashing throughout the archipelago, scientific collecting continued into the early part of the twentieth century. Various expeditions continued to remove tortoises, even as the scientists surely must have realized how few remained. Expeditions to Pinzón (then called Duncan) in 1897,

1898, 1900, and 1901 collected what were described as "last survivors." But they weren't. In an expedition in 1905–1906, eighty-six tortoises were discovered. All, including more than sixty females, were collected. Nonetheless, the population survived these human depredations and continues to survive on Pinzón today.

Not so fortunate was the Pinta population. The only known member of that island race is a single male, Lonesome George, who resides at the Charles Darwin Research Station. No Pinta females are known, so it is easy to see how George came by his appellation. A reward of $10,000 was offered to anyone who could procure a female Pinta tortoise but, whimsical Jurassic Park technology notwithstanding, money does not reverse extinction. No female was found on Pinta, and George remained lonesome, even rejecting female saddlebacks from nearby Isabela. Recently the reason for George's coyness may have been identified. George may, in fact, not be from Pinta. He may have been left there. No one knows how he got there. But recent studies on George's DNA indicate that he is from either San Cristóbal or Española, both of which are about 180 miles from Pinta. Fate may eventually be kind to George, should the scientists at the Charles Darwin Research Station decide to attempt breeding him with a female from Española or San Cristóbal. To my knowledge, that decision has not been made as yet.

Once the archipelago became a national park, earnest conservation efforts began. The Española race was believed to consist of but fourteen animals, two males and twelve females. Given the size of Española, meetings among males and females were so infrequent that human intervention was necessary to ensure that the subspecies would not become extinct from lack of contact with one another. Beginning in 1963, and lasting through 1974, the animals were captured and moved to the Charles Darwin Research Station, where it was hoped they would breed. The tortoises apparently found their new quarters agreeable, and procreation ensued. Their babies were kept at the research station for three years to permit them to grow to a safe size, at which time they were reintroduced to Española. The research station estimates that by 1995 there were seven hundred tortoises back on Española (none of which are near the visitor trails, so don't get your hopes up of seeing any). Similar efforts with other subspecies have also yielded encouraging results. On March 24, 2000, the research station celebrated the release of the one thousandth tortoise on Española.

In addition to safe rearing of tortoises at the Charles Darwin Research Station, efforts continue to eradicate species such as goats. When I last visited the islands in June of 1997, Volcán Alcedo was not open to tourists because there was an active program of goat killing. Alcedo, you may recall, harbors the largest single population of tortoises on the islands, and goat control is essential if this population is to continue to thrive.

The most recent threat to tortoise conservation appeared in the summer of 1996, when dead and sick tortoises were discovered on Santa Cruz near farms where visitors commonly go to observe free-ranging tortoises. Altogether, nine dead and eleven sick animals were discovered in a limited area known as El Chato. Examinations of the animals, their blood, and their feces have failed to pin down a cause, though a variety of bacteria and fungi are present, some of which (or none of which) may be pathogenic. One curious observation is that the stricken animals fed almost exclusively on passionflower fruit *(Passiflora edulis)* and that perhaps there is some chemical in the plants that is poisoning the tortoises. Even herbicide or pesticide poisoning has been suggested (the animals range over farmland), but no clear cause for the problem has yet been determined. Research continues.

The conservation of the Galápagos giant tortoise, like that of any species returned from the brink of oblivion, is an ongoing process. The efforts to date have been tolerably encouraging. Ten of the remaining eleven subspecies exist in reasonably viable populations on the islands upon which they evolved. In all likelihood, the Pinta race will die with Lonesome George. Three other races have also perished. But if the islands can continue to enjoy complete and total protection, and if eradication efforts against goats and other threats to the tortoises succeed, the hefty icons of the archipelago may once again become numerous, innocently greeting human visitors, gazing passively into the lenses of the cameras and camcorders, and enjoying a prosperity and security denied them for nearly three centuries.

Selected References

Beebe, W. [1924] 1988. *Galápagos: World's End*. Reprint, New York: Dover Publications.

de Vries, T. 1984. The giant tortoises: A natural history disturbed by man. In *Key Environments: Galápagos,* edited by R. Perry. Oxford: Pergamon Press.

Fritts, T. H. 1984. Evolutionary divergence of giant tortoises in Galápagos. *Biological Journal of the Linnean Society* 21:165–76.

MacFarland, C. G. 1972. Goliaths of the Galápagos. *National Geographic Magazine* 142:632–49.

Patton, J. L. 1984. Genetical processes in the Galápagos. *Biological Journal of the Linnean Society* 21:97–111.

Rick, C. M. and R. I. Bowman. 1961. Galápagos tomatoes and tortoises. *Evolution* 15:407–17.

Thornton, I. 1971. *Darwin's Islands*. Garden City, N.Y.: Natural History Press.

Tierney, J. 1985. Lonesome George of the Galápagos. *Science* 85:650–61.

Panorama of cliff side, Española

Orca with storm petrels and a
frigatebird, Bolivar Channel

BRUCE HALLETT

Palo santo forest, Floreana

BRUCE HALLETT

Waved albatross on its nest

Sea lions on beach, Santa Fé

Sea lion and pup, Santiago

BRUCE HALLETT

Male vermilion flycatcher

Marine iguanas sunning, Fernandina

Sally Lightfoot crab

Espumilla Beach at James Bay, Santiago, with tuff cone

Spatter cones near the summit of Bartolomé

Lava substrate with species of *Tiquilia,* Bartolomé

Tree-sized species of *Opuntia* cacti, Santa Fé

Lava field on Santiago

The genus *Miconia*, San Cristóbal

Flamingos parading, Floreana

Male magnificent frigatebird

Blue-footed booby pair billing

Red-footed booby, brown phase

Medium ground finch, Santa Cruz

Masked booby pair displaying

American oystercatcher

Galápagos mockingbird, Santa Cruz

View of Pinnacle Rock from summit of Bartolomé. Santiago Island is in the background.

Brachycereus species of cacti, Fernandina

Jasminocereus species of cacti, Santa Cruz

Forest of *Scalesia* species, humid zone at the Twins craters (Los Gemelos), Santa Cruz

Inside the humid forest, Santa Cruz

Mixed species of *Scalesia* and *Miconia* near El Junco Lake, San Cristóbal

Hood (Española) mockingbird

Swallow-tailed gull

Flightless cormorant

Large ground finch, Genovesa

Yellow warbler, San Cristóbal

Large cactus finch, Española

Marine iguana in surf, with Sally Lightfoot crab

Marine iguana with lava lizard sunning on its head

Marine iguana, typical coloration

Marine iguana on Española. Note its intense reddish sides.

Land iguana, Santa Fé

BRUCE HALLETT

Portrait of a marine iguana

Red-footed booby, white phase

Galápagos penguin

Galápagos dove

Male great frigatebird with throat pouch expanded

Lava heron

Short-eared owl

Dome-shelled giant tortoise

BRUCE HALLETT

Warbler finch, Santa Cruz

Saddleback giant tortoise

7

Hideous-Looking Creatures

There are few places on Earth where reptiles are essentially the dominant vertebrates of an ecosystem. Snakes and lizards and turtles, for the most part, mingle relatively inconspicuously among the far more obvious mammals and birds that populate terrestrial ecosystems. Crocodilians, in some sense relics of the Age of Reptiles, persist in numbers along some tropical riverine areas. And on the small Indonesian island of Komodo, one among many of the Lesser Sunda Islands, it is still possible to find large carnivorous lizards fully capable of running down and killing deer, the Komodo dragon lizards. These 9-foot (2.7 m) monitor lizards inspire some zoologists to study them as contemporary models of how a dinosaur species such as *Tyrannosaurus rex* might have functioned in its Cretaceous ecosystem. And then there is the Galápagos archipelago, where you can find not only giant tortoises but also two kinds of larger than average lizards. And these hefty saurians do look just a bit like latter-day dinosaurs. The Galápagos Islands can still justify the title of "realm of the reptiles." And for many among us humans, reptiles are an acquired taste. Darwin called the marine iguanas "hideous looking creatures," a description also applied to the animals by Captain FitzRoy. Hideous? Look at them, learn about them, and you be the judge.

Marine Iguanas

The resemblance, such as it is, between a lizard, no matter how grand its size may be, and a dinosaur is strictly superficial. Dinosaurs, in spite of their name that means "terrible lizards," were not lizards. Dinosaurs were unique, differing from lizards in several important anatomical respects. But lizards, especially large ones, do superficially resemble reconstructions of dinosaurs and thus have a kind of Mesozoic aura about them. And an appropriate background setting helps. Thus, when you first set foot on almost any of the Galápagos Islands, it is not unusual for your imagination to backtrack many eons, to the Age of Reptiles. Such musings are kindled by the presence, often in substantial numbers, of scaly-skinned, beady-eyed, salt-encrusted black marine iguanas.

Marine iguanas get your attention. The biggest of their lot approach 4.5 feet in length (about 1.33 meters), tails included, and can weigh close to 20 pounds (9 kg). Most are black, but with some green as well as bright pink, even red, along their sides. There is variability in iguana coloration among islands, with the darkest and largest marine iguanas tenanting Bolivar Channel, between Isabella and Fernandina, and the brightest, most reddish of the lot found along the cliffs of Española.

Wherever they occur, the iguanas, like other Galápagos animals, stare rather blankly and show no fear, and thus the human visitor can walk among them, over them, and around them. They generally stay put, staring with utter nonchalance, some sprawled on ragged black boulders, some resting atop others of their kind. They show only indifference as a Darwin's finch plucks a parasite from their skin or Sally Lightfoot crabs scamper restlessly up and over the lounging lizards. Some marine iguanas occasionally bear living "hood ornaments," much smaller lava lizards that habitually perch atop their heads, their larger cousins showing no sense of intrusion or giving the slightest hint of knowing how ridiculous they look to the human visitor. Now and then the robust black lizards move, a kind of reluctant shuffle, perhaps to change position or to meander to the water's edge to feed.

Their faces are shortened and rounded, almost puglike, and they have a row of small flexible spikes running along their backs, the spikes increasing in length as they trail up over their napes. A back lined with spikes, even little ones like these, goes a long way to providing a dragonlike countenance. If they were larger, they'd be really imposing. The lizards' sprawling legs give them a decidedly awkward gait. To some human eyes, these creatures pass muster as miniature versions of Godzilla. To William Beebe, two wrestling marine iguanas, backlit by the sun, were reminiscent of a painting of a species of *Tyrannosaurus* that he recalled, inspiring him to comment that he "had the conviction of ancient days come once more to earth."

Marine iguanas *(Amblyrhynchus cristatus)* are neither mini-Godzillas nor tiny *T. rexes.* Sufficiently interesting in their own right and endemic to the Galápagos Islands, they represent the only species of truly marine lizard. You won't find them anywhere else on the planet, nor will you encounter any lizard quite like them. They abound throughout the islands, with particularly notable populations on Española and Fernandina Their total population now numbers in the hundreds of thousands. Estimates vary but 250,000 total is reasonable.

It is, of course, not known when the first immigrant iguanas arrived in the Galápagos Archipelago, presumably refugees from somewhere in South or Central America. Nor is the ancestral species known with any certainty, though the abundant and widespread common iguana *(Iguana iguana)* is certainly a reasonable possibility as the grand ancestor. The original colonization event could have occurred several million years ago, though one estimate puts it as recently as ten thousand years ago. The animals could have even colonized much more recently. In any event, evolution can take place rapidly, especially in small isolated populations, and apparently it has, to adapt these unique creatures to their current lifestyle. How could a rainforest lizard be transported to a distant archipelago?

Visitors to the Galápagos often fly to and from Guayaquil, Ecuador, and if the visit to the islands should coincide with the rainy season, the swollen Guayas River flow-

Marine iguanas sunning, Fernandina

ing past Guayaquil will be readily visible from the air. Noteworthy are the myriads of small islands of floating vegetation, mats broken from the banks by the river's energy, some of substantial size, little islands of jungle, now being transported seaward to become waifs of the Humboldt Current. Common iguanas, abundant throughout the lowland tropics, are partial to riverine trees. The iguanas often abandon branches high above to leap to safety in the river below should harm threaten them. These large herbivorous lizards are impressive swimmers, a behavior that adapts them to cope with predation threats as well as allows them to swim across a river or stream, even one rather wide, merely to get to the other side. Thus it is far from implausible that the ancestor of all the marine and, for that matter, land iguanas on the Galápagos might have been a pregnant common iguana trapped on a floating mass of vegetation torn from the river border in equatorial South America and destined for a journey partway across an ocean.

Reptiles are excellent island colonists. Their poikilothermic (crudely termed *cold-blooded*) metabolism is far less demanding of food than the homiothermic *(warm-blooded)* demands placed on birds and mammals. A snake's heart may beat as many times in a day as a hummingbird's does in a few minutes. A reptile can fast for a far more protracted period of time than a mammal of equivalent weight. Reptiles have dry skins resistant to water loss. Thus a common iguana could endure a prolonged period without fresh water, tolerate daily exposure to intense sunlight, be deprived of food, yet still survive a long oceanic transit between South America and the archipelago. It would not necessarily be a pleasant trip, but it would be survivable. It remains to be shown, however, if the molecular genetics of the common iguana and the marine iguana suggests that the latter is derived from the former.

If the scenario outlined in the previous paragraph is true, or at least somewhat true, it must also be true that some of the original iguana colonists never really left the shoreline. Marine iguanas do not venture far inland but essentially spend their entire lives within the intertidal zone and the sea immediately around it. Marine iguanas do not look like common iguanas, natural selection having shaped them adaptively within the context of their peculiar lifestyle.

Like common iguanas, marine iguanas are vegetarians, but unlike their close cousins, the land iguanas, they completely reject terrestrial vegetation and dine instead almost exclusively on nutrient-rich marine algae growing subtidally and within the intertidal zone. The marine iguanas' shortened faces adapt them well for nipping algal tufts from rock faces, rather like sheep close cropping grasses. To the untrained eye, marine iguanas appear to be "kissing" bare rocks as they turn their heads downward and push their oddly blunted faces directly against the rocks. But the rocks are not bare. Close inspection reveals numerous tufts of mostly green algae, the main food of the iguanas. In many places throughout the archipelago, the algae look unimpressive, hardly sufficient to maintain thousands of large hungry lizards, even given the reptiles' generally low metabolisms. But algae, not only on the Galápagos but elsewhere in the world, grow with uncommon speed, especially when immersed in nutrient-rich seawater. Were it not for the iguanas, algal mats would densely cover the rocks, and in some places, even with iguanas, the rocks are densely covered. But, much as greens keepers with lawn mowers encourage the growth of putting greens by keeping them well cropped, the constant attention of hundreds of foraging iguanas tends to keep the algae both unusually short and growing at a maximum rate.

The lesson here is that the rate of growth, the biomass added *per unit time,* is, in fact, more important than the total amount of growth present at any given time. A seemingly light coating of algae can, if it grows rapidly enough, sustain a very large mass of iguanas. Weigh all the iguanas feeding on a section of intertidal zone, and then weigh all of the algae found there, and the result will often be that the animals appear to significantly outweigh the plants, in seeming violation of both common sense and the second law of thermodynamics. However, the amount of energy that flows *per unit time* from the sun through the algae is vastly greater than the energy that flows per unit time through the iguana population (just as humans outweigh all the burgers in a crowded fast food restaurant—but the burgers keep coming, faster than the humans, in fact).

Unless you attain an understanding of the feeding ecology of marine iguanas, the interesting story behind the seemingly sparse algal growth throughout much of the Galápagos would likely go unnoticed. Given their dietary preference, the puglike face of the lizards becomes understandable as an adaptation. It permits the animal to make wide contact with the substrate and obtain the most algae for its nipping efforts.

Not surprisingly, algal growth is generally richer subtidally, where constant immersion allows more time for the plants to grow but where the lizards must swim farther to eat. The largest of the marine iguanas, the big males, tend to be most common in the deeper waters. Females and juveniles are less apt to move out of the in-

Marine iguana scraping algae from a rock surface, a common foraging behavior

tertidal zone. To a certain extent this segregation is due to physical realities. Waves crash hard on Galápagos rocks, currents are strong, and only the largest, most robust animals can maneuver through the often-heavy surf. It would be of obvious danger to an iguana to become captured by a powerful current and washed away from the islands. The iguanas are strong, but the Pacific Ocean is a whole lot stronger. So it is the largest, strongest animals that do best at breaching the heavy surf and swimming out to the deeper algal-rich areas.

The body size of marine iguanas varies with food abundance around the islands. A study conducted by researchers Martin Wikelski and Fritz Trillmich compared marine iguanas on Santa Fé, where the animals are typically large, with those on Genovesa, where they are considerably smaller (weighing a maximum of 2 pounds [900 g] compared with 8 pounds [3,500 g] on Santa Fé). Food availability differs between the two islands, with Santa Fé having the richer feeding grounds. More food means bigger iguanas. But iguana body size is influenced by more than food availability.

Marine iguanas fare badly in years of severe El Niños, when the upwelling system that provides essential nutrients to the algae is disrupted by the warming of the sea. Inevitably the lizards suffer significantly higher mortality rates, with the largest animals suffering the highest mortality rates. Large animals need higher quantities of food, and the food is simply not available during the worst times of an El Niño; thus the big animals succumb.

However, researchers Martin Wikelski and Corinna Thom learned that marine iguanas do have an amazing short-term response to food limitations imposed by El Niños: They shrink! In general, animals get bigger as they get older until they reach

Marine iguana swimming

adult size, when they essentially stop growing. But two studies, one eighteen years in duration and the other eight years, have shown that marine iguanas may shrink their bodies (from snout to vent) by as much as 20 percent over a two-year period. The shrinkage correlates with El Niños, times of food shortage, and reverses during La Niñas, when upwelling resumes and food becomes much more abundant. Large iguanas shrink proportionally more than small iguanas, and females shrink more than males. The overall shrinkage is sufficiently great that it must be accounted for by actual bone absorption, not merely decreases in cartilage and connective tissue. The studies showed that those individuals that shrunk the most survived the best. The mechanism for how the shrinkage actually occurs is still under study, but it appears related to hormonal changes, especially corticosterone levels. Individual iguanas are capable of shrinking and lengthening several times in their relatively long lives, as El Niños come and go. Body length variation is a short-term adaptive response to a stressful condition and seems to represent the only known case of a vertebrate animal changing size routinely during its adulthood.

Marine iguanas are excellent swimmers, a behavior that, as mentioned above, is characteristic of their (possible) ancestors from mainland South America. But marine iguanas have evolved anatomical and physiological adaptations unique to their lifestyle. Their tails are laterally flattened (from side to side, like that of a crocodile), making it easier for the tails to scull through the water. When marine iguanas swim, they swim like crocodilians, their legs held tightly against their sides, the force of movement being provided mostly by the sinuous undulations of their long flattened tails. Many Galápagos visitors enjoy snorkeling while marine iguanas swim past and dive down to forage on the algae-covered rocks below. Most marine iguanas are just as fearless of people when in the water as they are when on land. The iguanas' long claws help them hold fast to the rocks as they forage underwater.

The dark coloration of the marine iguanas (many are almost solidly black) gives them a certain camouflage against the dark rocks upon which they congregate. But cryptic coloration, which may be highly adaptive for juveniles, is likely not of major importance for adults. The adults are relatively free of predation from land creatures, though some fall prey to Galápagos hawks. Instead the dark coloration of the skin probably functions mainly to absorb solar radiation to heat the animal as quickly as possible. Put simply, dark iguanas heat up more quickly than light iguanas, and heating up may be adaptive.

Recall that these animals are quite unable to warm themselves physiologically. They must depend on the sun in a very direct way. Galápagos waters abound in nutrients but are generally cold. Even animals feeding in the intertidal zone can chill quickly when wet. Dark skin allows for maximum rates of reheating. Indeed, marine iguanas are known to bask for a period of time before entering the sea to feed. Measurements suggest that the animals attain a body temperature nearly that of a human, 95°F (35.5°C). Human body temperature is 99°F (38°C). The animals' temperatures can, on occasion, rise to nearly 104°F (40°C). At that point, the lizards must pant or move quickly to the water to lose heat or they will suffer the consequences of overheating.

As marine iguanas forage, they constantly lose body heat (water is a great conductor of heat as anyone knows who has gotten cold from merely remaining too long in a bathtub), and this is likely a factor in determining the length of each foraging excursion. It is also noteworthy that marine iguanas swim relatively slowly, a characteristic observed by many snorkelers. Such behavior may serve to conserve energy, given that the object of the animal's quest, algae, is not going to swim away. When a chilled iguana leaves the water, it will lie spread-eagled on warm rocks, where it can absorb heat by conduction and bring its body temperature up.

Careful studies on the physiology of marine iguanas reveal that they fatigue rapidly, more rapidly, in fact, than a mammal would. Such a pattern is typically reptilian, a characteristic of animals that utilize primarily anaerobic respiration, generating lactic acid as a product of their breakdown of glucose. Mammals do this as well—it's why humans typically develop sore leg muscles if they overexert by running too much—but marine iguanas rely on anaerobic metabolism for 97 percent of their use of glucose, far more than mammals do. Such a pattern is not atypical of terrestrial lizards, and because of such a similarity, it has been argued that marine iguanas must have evolved from a terrestrial species, hardly an earth-shaking suggestion. More importantly, anaerobic metabolism is much less oxygen expensive, permitting the animals to remain submerged without surfacing to breathe. It also permits them to remain active at a cooler overall body temperature than a mammal of equivalent size and weight. Darwin provided an account of how he and his shipmates weighed down a marine iguana, tossed it overboard attached to a rope, and retrieved it a half hour later. It was fine.

Marine creatures normally have little or no access to fresh water and thus adapt to drinking brine. Marine iguanas do this well, with the aid of salt glands positioned unobtrusively above each eye and directly connected by a duct to the nostril. The glands concentrate salt and dilute the ingested seawater to metabolically tolerable concentrations, and once a certain amount of brine is collected in the glands, it is passed along to the nostrils and quite literally sneezed out. The moisture from the iguanas' mucus evaporates, and the crystallized salt remains behind. Consequently, the animals typically have encrustations of salt on their heads, a kind of cephalic margaritalike decoration. Salt glands are common on many species of marine birds as well.

Marine iguanas are basically colonial, a characteristic that may be a byproduct of their relatively limited habitat, the intertidal and subtidal zones. Coloniality is anything but commonplace among lizards. But living within such a narrow band of habitat would tend, naturally, to concentrate the animals. They live in apparent peace with one another much of the time and routinely bunch together, perhaps to retain heat. Tension among the animals happens during the breeding season, which varies somewhat among the islands. Not surprisingly, it is the males that become the most tense.

In the wide spectrum of mating rituals in the animal kingdom, some species exhibit male rivalry to such an extent that the animals may become seriously injured, sometimes fatally so. Scarred sea lions, bearing the marks of years of male/male combat, are the norm for various sea lion species, including those inhabiting the Galápagos. But in most animal species, males do not really fight, opting instead for an approach more typical of so-called professional wrestling. Male iguanas that wish to propagate their genes congregate in loose associations termed *leks*. The contests appear more dangerous than they really are. The loser animals are afforded an opportunity to retreat with their flesh intact. Male marine iguanas seemingly dress for their territorial encounters, their pigmentation intensifying as part of their hormonal preparation for launching the next marine iguana generation. On most islands the males become much more intensely pigmented, usually with red, during the breeding season. Given that reptiles see colors, it is well within the realm of possibility that redder is better as far as potential for intimidation goes. Or perhaps females like red a lot. I don't think anyone knows. But the males do become rather intensely red, especially along their flanks. Big males have thickly scaled heads, with tubercles protruding among the encrusted salt that continuously adorns their crania.

When males confront one another, they posture, bob heads (a common behavior among male lizards of many species), and, somewhat like saurian goats, whack their skulls together as they try to shove each other. This is the behavior that so impressed Beebe, making him imagine that he was watching dinosaurs. Presumably the well-protected heads of the iguanas can endure the butting without serious detriment to the creatures. No blood is usually drawn, and eventually one of the two combatants will break off, with a winner and a loser. It is not known if the animals suffer from postcombative headaches, and it wouldn't matter anyway because natural selection will be obeyed.

One reason why male marine iguanas are larger than females may be because their large size aids in driving away other males and so favors access to females. Thus the forces of sex drive male marine iguanas to evolve larger body size, putting these animals at most risk during an El Niño event (see above), a case of what evolutionists call "opposing selection pressures."

It is not known what the variation in mating success is among male marine iguanas. Perhaps only a few of the males in a colony typically mate, the dominant males. Or maybe most mate, but the quality of females varies, with the most dominant males able to select the highest-quality females. All of the intimate details of iguana society are not yet known. What is obvious is that males must entice seemingly reluctant females to mate, an utterly normal behavior among animals throughout the wide world of biology. The male carries the "cheap" gamete, the sperm, easily replaced if wasted. The female carries the "expensive" gamete, the egg, not so easily replaced given the much greater amount of energy that goes into the making of an egg cell compared with a sperm cell. Further, once copulation is completed, males have no other responsibility in the process, whereas females now have the males' genes invested with their own in eggs that must mature and be laid. In each mating season a given male may mate with numerous females, but females, given the reality of nesting, are far more limited in their choices. Thus it should be surprising to no one that it tends to be female animals that are coy, are reluctant to mate, and employ various methods to assess the males before making the commitment to mate. Males, on the other hand, are anything but coy, attempting to mate as often and with as many females as possible, all because sperm are little and eggs are big.

In the world of marine iguanas, a male will posture to a female, the courting male making slow and deliberate circles around the female, all the while head bobbing in typical lizard fashion. Females often appear anything but eager to engage in sex. Often they appear oblivious. Because of the aloof nature of the females, it becomes apparent why male ritualistic combat is so much a part of iguana mating behavior. If a male fails to secure a territory in which to court females, he will be unable to pursue the opposite sex without interruption by other males, and pursue he must.

For whatever reason known only to her, a female iguana will eventually accept a suitor, and the male will mount by lumbering up on her, grasping her neck while manipulating his tail beneath her such that his organ of intromission may reach its destination, the female's vent. This procedure looks awkward, but it works. Male lizards (and snakes) have hemipenes, a paired penislike organ. One of the two hemipenes is used in each copulatory event. The organ has sinuses that fill with blood and cause the hemipene to become erect, allowing it to pop into the female's vent. The tip of the organ has small spines that aid in keeping it in place until the semen is delivered. The semen is actually ejaculated into a space called the cloaca (from the Greek word for "sewer"), a common vestibule for the products of the urinary, intestinal, and reproductive system (though not at the same time). From the cloaca, sperm presumably move to the correct duct and fertilize the female's eggs. Once a male has mated, his bright coloration quickly fades until his hormonal cycle reboots for the next breeding season.

Females lay eggs in shallow burrows located in flat, sandy areas, some of which are roped off on various islands to ensure that human visitors do not disrupt the eggs. Females often remain near the nests to drive off other females until most other females have laid their eggs. This behavior is adaptive in that it prevents another female from inadvertently digging up the eggs and thus destroying the clutch. It is not known if nesting sites are so limited in area as to produce serious competition among females for access to prime sites. Normal clutch size varies from one to four eggs and, typical of reptiles, the eggs possess a leathery shell quite different from that of birds' eggs. The eggs develop without incubation by females (another normal reptilian characteristic that sets reptiles quite apart from birds), and it requires between three and four months under the equatorial sun for the eggs to hatch.

Baby iguanas, of course, never know their parents, both of whom are anonymous by the time the eggs hatch, even if still resident within the same colony. Hatchling iguanas are rather lizardlike in appearance, having not yet developed the distinctive shape of the adults. They are also variegated in patterning, with a kind of marbled look. Hatching is a time of high mortality for marine iguanas. Herons, hawks, lava gulls, and feral cats have all been observed preying on young marine iguanas. It is at this stage of the life cycle when cryptic coloration is probably of greatest selective advantage.

Land Iguanas

Darwin thought land iguanas were "stupid" in appearance, but I think that even he, were he still around, would upon reflection think that comment to be unjust. He was, after all, homesick at the time he wrote it. Land iguanas appear neither stupid nor perceptive, their quiet, admittedly unexpressive, red eyes making them look merely indifferent. I have yet to meet a reptile that looks otherwise. They do look impressive, however. Land iguanas, like their marine brethren, reach lengths in excess of 3.28 feet (1 m). Their yellow heads are characterized by numerous, well-defined, deeply ridged large scales, vastly larger than those covering their bodies. They have more elongate faces, more snoutlike, than their marine relatives. A ridge of flexible spines runs high along the nape, grading to shorter spines along the dorsal surface of the back. A large flap of skin hangs limply below the throat. Overall the color of these robust creatures is a pale yellow above and brown below, strikingly different from the blackish marine iguanas. There is no mistaking the two.

To the average tourist, marine iguanas will seem far more abundant than land iguanas, and indeed, they are. Vast hordes of land iguanas once roamed several of the islands, but their numbers have been profoundly reduced, and they have been totally eliminated from at least one island where once they lived in great abundance. Charles Darwin commented on the immense number of land iguana burrows he observed on James (Santiago) Island and complained that the burrows were so dense that there was no room to pitch a tent easily. Land iguanas aren't there today. Like giant tortoises, land iguanas may have been persistent victims of rats, pigs, dogs, and humans, as well as losers in competition with feral goats. Feral dogs were blamed for the extirpation from Santiago, but this, itself, is a mystery because there are now no feral dogs on that island and have not been for some time. So nobody really knows what happened to the iguanas.

Land iguanas continue to thrive on parts of Fernandina, though tourists do not see them there because they are in the caldera, far inland from Punta Espinosa, where tour pangas land. They also occur on some areas of Isabela and Santa Cruz. But it is on the little island of Plaza Sur (South Plaza) where the Galápagos land iguana is seen by most tourists. Some three hundred or more inhabit an island so small (only 13 acres, or about 5 ha) that it can be circumnavigated on foot in a matter of an hour or so. Indeed, after you gingerly maneuver around the lounging mass of sea lions that normally congregates on the small dock at Plaza Sur, it is the land

Land iguana, Plaza Sur

iguana that most immediately captures your attention and film. Usually there will be several in view at once, ideal photo opportunities, their yellowish bodies wide and fat as they sprawl beneath tree-sized cacti of the genus *Opuntia* or graze purposefully on the blossoms of the genus *Portulaca*, which serves as a most attractive ground cover over much of the island.

Elsewhere among the islands the land iguana can be seen at Santa Fé, where the animals are paler and more uniformly yellow. This population is usually considered a separate species *(Conolophus pallidus)* from the more widespread *C. subcristatus*. Unlike the ease with which the creatures are encountered on South Plaza, those on Santa Fé are much more difficult to find. Santa Fé supports a vast forest of the tree-sized *Opuntia* species of cacti, among the most striking examples of this remarkable species on the islands. In addition there is, especially during the wet season, a thick ground cover of various plant species. Unlike Plaza Sur, on Santa Fé it is easy for the iguanas to remain cryptic in the dense ground cover. The number of animals per acre is far less, which makes detection much more challenging.

There are two ways to find land iguanas on Santa Fé, the hard way and the easy way. The easy way is to take a short trail from the beach and hope you get lucky. You probably won't. When I walked this trail with my group, we found only a baby iguana, and there was considerable discussion among the guides as to whether or not this diminutive creature was a land or a marine iguana. The consensus finally came down on the side of marine iguana. Later on the trail we found a larger juvenile iguana, this one clearly a land iguana, but a juvenile nonetheless. We wanted to see an adult. So we found it the hard way. The hard way is to walk a much longer

Pale land iguana, Santa Fé

trail up a rocky escarpment into a flattened tableland where land iguanas become considerably more abundant. The trail is steep, the walk hot, the pace slow. Once there, we looked carefully among the brush until something with a reptilian countenance looked back at us. That was the land iguana. The Santa Fé land iguana lives within the confines of a splendid forest of the *Opuntia* species with a dense understory of grasses along with some palo santo and *Scalesia* species, as well as numerous other shrublike plant species. The panoramic view from the escarpment is among the best on the islands.

Land iguanas are primarily vegetarians, feeding mostly on the *Opuntia* species, spines and all. They manipulate and remove spines using their claws but still must encounter many spines while chewing, swallowing, and digesting—a tough salad indeed. But the pads of the *Opuntia* species are a nutritious food source and also serve to supply sufficient water during the dry times of year. Land iguanas, like giant tortoises, will drink profusely when given the opportunity but can go long periods without drinking as long as they have access to the *Opuntia* species. Other plants are eaten as well, the diets varying somewhat among islands. Fruits and flowers are eaten as well as grasses. Land iguanas are also known to feed on carrion, sea lion afterbirths, insects such as grasshoppers, and even centipedes (which can give a very powerful bite but, then again, these iguanas are used to cactus spines).

Land iguanas are thought to be colonial, at least in a loose sense. Obviously the ones on Plaza Sur are colonial because they have no place to go and there are a lot of them! Elsewhere they congregate in the caldera on Fernandina and on the tableland on Santa Fé.

Mating is probably harder on males than females. Males must earn territories through direct combat, a bit more direct than is the case with marine iguanas. Males posture aggressively at one another, performing such actions as arching their backs and swelling their bodies, all perhaps to make the animals look larger and scarier than they really are. Blood is occasionally drawn as the creatures strenuously butt heads, a male concession to giving up good sense in favor of good genes. Territories vary from about 33 to 65 feet (10 to 20 m) in diameter, a lot of real estate for one iguana to manage.

Male land iguanas are substantially larger, heavier, and more colorful than females, characteristics that strongly suggest sexual selection through male/male competition. Sea lions have some of these same characteristics. In addition, male land iguanas have much larger spines on their necks than females do, perhaps an adaptation either to intimidate rival males or to impress females, who ultimately choose with whom to mate.

Courtship is somewhat similar to that observed for marine iguanas, with males bobbing, circling, and hoping (or whatever passes for a hope within the presumably narrow confines of reptilian consciousness). Copulation is essentially the same as for marine iguanas. It is reported that a successful male may mate with up to seven females, each female laying her eggs in the nesting territory defended by this dominant male. Again, this characteristic signifies sexual selection, where some males are big genetic winners and others are total losers, unable to reproduce for lack of a territory.

Female land iguanas excavate nesting burrows. They dig several before actually laying eggs in one. Female competition for nesting sites has been reported. Indeed, a female will remain within her nesting burrow for about a day or so after laying eggs, will cover the burrow when she leaves it, and will remain and defend the burrow from intruders for about a week or so. As with marine iguanas, three to four months are required for the baby iguanas to hatch. Clutches are larger than marine iguanas' and range from a half dozen (sometimes fewer) to as many as two dozen or more eggs. Clutch size differs among various island populations, probably due to food availability (to make eggs) or perhaps genetic factors evolving in relation to infant mortality, which is reported to be very high. Survival rate among baby land iguanas is less than 10 percent, mostly due to predation by hawks, owls, snakes, and herons. Eggs can fall prey to beetles. It would be interesting to examine the correlation between average clutch size and infant mortality among various populations.

The daily life of a land iguana is probably less than riveting. Adults have no natural predators, so they may come and go pretty much as they please. Like the rest of lizardom, the animals spend much time basking in the sun, heating themselves in the morning to a body temperature that ranges from 90°F (32.2°C) in the dry season to 97°F (36°C) in the wet season. In the heat of the day the iguanas typically retreat to the shade. Both sexes spend the night in burrows, presumably to preserve body heat—they fear little from predators.

Land iguanas control small predators, such as ticks and mites, by permitting small and medium ground finches, as well as mockingbirds, to hop atop them, under them, and beside them so that they can pluck ectoparasites from the iguanas' scaly skin. To do this, the iguanas often raise themselves up, appearing to stretch, thus affording a bird access to pretty much everywhere on them. This mutualism provides the bird with food and eliminates the offending parasite from the iguana, a nice deal for both parties.

A life of eating, sunbathing, hiding out in the shade, bird-facilitated grooming, and occasional mating can last a land iguana up to seventy years.

To return to a subject explored earlier in the chapter, no one knows when the first colonist iguanas reached the Galápagos or precisely which species they were. Further, it is not clear if land iguanas have separate ancestors from marine iguanas, either in the form of different ancestral species or the same species but two separate colonizations. Some researchers estimate a mere ten thousand years since colonization; others suggest a longer period. At any rate, as with Darwin's finches and the mockingbirds, genetic divergence and speciation have occurred. Anatomical studies suggest a close similarity between marine and land iguanas, so they may have emerged from the same ancestral species. However, anatomical studies can sometimes be deceiving.

Genetic differences exist among the different populations of land iguanas. This is not surprising given that they are generally sessile, without much occasion to travel among the islands, and thus the isolated populations would be expected to diverge due both to genetic drift and natural selection. The fact that the Santa Fé population is regarded by most authorities as a separate species is really not surprising even if it has been only one hundred decades since the first colonization.

Land iguanas have become a focus of conservation research at the Charles Darwin Research Station on Santa Cruz. Like the eggs of tortoises, eggs of land iguanas have been taken from threatened populations such as those at Conway Bay on Santa Cruz and Cartago Bay on Isabela and incubated at the research station. The young are raised for either captive breeding or reintroduction. Animals are now being reintroduced, including in areas such as Baltra, where former populations were totally extirpated.

Lava Lizards

While marine and land iguanas tend to command center stage, there is a scaly supporting cast. There are seven species of lava lizard, all endemic, which occur in some combination or another on almost all of the islands. There are also nine species of geckos (another kind of lizard), six of which are endemic, and three snake species, each endemic. Of these various animals, it is the lava lizards that are most often encountered.

Lava lizards, all in the genus *Tropidurus*, probably deserve more wonder from tourists than they get. Like Darwin's finches they have undergone a fascinating evolution and make good subjects for the ongoing study of how evolution affects the genetics of populations. They are certainly obvious and attractive little lizards, ranging in size up to somewhat over 10 inches (25 cm). Size varies among islands, and the largest of the lot is on Española. They all look like typical lizards, not nearly so striking as the wondrous iguanas. The female lava lizard is smaller than the male and has a bright orange throat, the orange extending up to cover some of the face. The male is pale tan, darker along the sides, and mottled with black speckles. There is much variability among islands and among species. Some of the variation seems to result in a cryptic pattern, probably attributable to natural selection imposed by predators. For example, the animals that inhabit dark lava are noticeably dark, while those found on light sandy beaches are pale, a fact easily documented by even the most casual observer. Lava lizards are not found in the highlands but rather inhabit the more arid areas of the islands. They are diurnal and easy to observe.

Most tourists at one time or another notice the curious push-up behavior so common among lava lizards. Both sexes do push-ups, which are exactly as they sound to be: The animal raises and lowers itself, up and down, up and down, all in very deliberate fashion. There are probably any number of home videos, filmed by tourists, that feature the bobbing lizards. Actually this behavior is not rare among lizards, but the Galápagos lava lizards actually seem to use push-ups as a kind of code among the various species. One study showed that the particular patterning of the push-ups varied remarkably among the island populations, even for the same species on different islands. The function of the display is to signal other lizards, usually regarding territory but also, in the case of opposite sexes, courtship as well. Lava lizards are strongly territorial (so there are lots of push-up opportunities), each sex defending its own territory.

When I was a graduate student, a friend of mine, also a graduate student, studied box turtles one summer in New Jersey. He made an enclosure, some of which was in the sun, some of which was shaded. He wired up box turtles with thermistor probes to monitor their body temperature, and he kept track of air temperature as well. He learned that when the sun is beating down, the turtles move into the shade. When it is cool, the turtles move out into the sun to warm up. By simply moving between sun and shade, the box turtles kept a body temperature far more constant than the ambient air temperature. This is the reptilian way, and lava lizards are also great practitioners of the art of controlling body temperature by behavior.

Early in the morning, when the air is cool, look for sunning lava lizards exposed on rocks. You might note that the animals expose themselves perpendicular to the sun's rays, the quickest way to gain heat. By the warmest part of the day they seek shade. Many visitors notice the curious gait of lava lizards over hot sandy areas or lava fields. These places get very hot (try walking over them with bare feet), and the lizards lift themselves up on their toes as they make their way over these surfaces.

It has been learned that lava lizards can live for a decade. Females become sexually mature at an earlier age than males, perhaps because male/male competition for access to females and territories is sufficiently severe to affect the evolution of sexual maturity in males. In any case, females lay eggs in excavated nests, and incubation, as with the iguanas, takes about three months. Baby lava lizards are undoubtedly subject to heavy predation, though clutch size is generally small, between three and six eggs. However, a given female will nest repeatedly within a year and many times throughout her life, so from many attempts comes at least some eventual success. Lava lizards are anything but rare.

Geckos

Most lizard species are diurnal, not an unexpected pattern considering their poikilothermic metabolism. But in tropical regions, where it stays tolerably warm at night, there is a kind of lizard that has evolved to be nocturnal. It is the gecko. There are many species of geckos throughout the world, and a good number of them are no strangers to humans, the geckos entering houses to feast on insect invaders, including the universally disliked cockroach family. Consequently geckos are generally welcomed and considered good luck. Geckos have another attribute, uncommon among lizards, which is that they are highly vocal. Geckos make a variety of chirps and tweets that range from charming to rather irritating, depending upon whether you are a light or heavy sleeper.

There are nine species of geckos on the Galápagos, six of which are in the genus *Phyllodactylus*. These animals have been longtime residents and are endemic to the archipelago. Three other species have recently been introduced.

Geckos have largish heads, a characteristic accompanying the fact that they have evolved large eyes, presumably in response to their nocturnal habits. Their other notable characteristic is their feet, which feature toes that are adorned with numerous tiny skin folds enabling the lizard to scamper effortlessly up a smooth vertical surface, like a wall, for example.

Because they only come out in the dark, and because most tourists remain on boats at night, Galápagos geckos are not often seen by visitors to the islands, but look for them if you should spend an evening at Puerto Ayora on Santa Cruz. They're around.

Snakes

If you should ever decide to visit the island of Guam in the South Pacific Ocean, you need not bother with binoculars and bird guide. There are no birds. The snakes got them. Some decades ago a snake species was introduced to the island. It multiplied, it climbed trees, it entered burrows, it found ground nests, and it fundamentally eliminated the entire terrestrial avifauna. Which is why it is curious that snakes do not, in fact, pose much of a threat on the Galápagos. In fact, on the Galápagos, the birds (Galápagos hawks) eat the snakes.

There are three species of snakes that inhabit the archipelago, and they are generally widespread, especially on the central and western islands. Visitors do not often see them because they are rather small (rarely exceeding 4 feet, or about 1.2 m) and tend to keep out of sight. The little snakes are attractive, each species being some combination of yellow and brown patterning. Each species kills its prey by constriction, coiling about the victim and preventing it from breathing until it succumbs. The usual prey ranges from rats to grasshoppers and includes such delicacies as lava lizards, baby iguanas, and geckos, as well as nestling mockingbirds and finches.

Galápagos snakes trace their probable ancestry to a mainland species found in coastal Chile and Peru. When it arrived on the islands is unknown. In any case the presence of the serpents has not resulted in an ecological cascade of destruction such as was experienced in Guam. For that we can all be grateful.

Selected References

Dawson, W. R., G. A. Bartholomew, and A. F. Bennett. 1977. A reappraisal of the aquatic specializations of the Galápagos marine iguana *(Amblyrhynchus cristatus). Evolution* 31:891–97.

Laurie, A. 1983. Marine iguanas on the Galápagos. *Oryx* 17:18–25.

Snell, H. L., H. M. Snell, and C. R. Tracy. 1984. Variation among populations of Galápagos land iguanas *(Conolophus):* Contrasts of phylogeny and ecology. *Biological Journal of the Linnean Society* 21:185–207.

Steadman, D. W. and S. Zousmer. 1988. *Galápagos: Discovery on Darwin's Islands*. Washington, D.C.: Smithsonian Institution Press.

White, F. N. 1973. Temperature in the Galápagos marine iguana: Insights into reptilian thermoregulation. *Comparative Biochemistry and Physiology* 45:503–13.

Wikelski, M. and C. Thom. 2000. Marine iguanas shrink to survive El Niño. *Nature* 403:37.

Wikelski, M. and F. Trillmich. 1997. Body size and sexual size dimorphism in marine iguanas fluctuate as a result of opposing natural and sexual selection: An island comparison. *Evolution* 51:922–36.

8

All the Birds Were So Tame

Reptiles have done it, way back in the distant Triassic and Jurassic periods. More recently, mammals have accomplished it, rather well, in fact. And birds have done it, though perhaps a bit less thoroughly than mammals and reptiles. What is "it"? Each of these classes of vertebrate animals is fundamentally terrestrial, adapted to a life on land, terra firma, where gravity constantly asserts itself and where each breath comes from oxygen contained among the gases of the atmosphere. But each class contains species that have effectively adapted to life in the oceans, albeit still breathing air but otherwise evolving to be as much a part of the ecology of the oceans as the many species of sharks and myriads of bony fish. In this chapter we look at seabirds of the Galápagos, a unique assemblage, some of which make the Galápagos Islands one of the most coveted places for bird-watchers to visit. Birders come from around the world, making the pilgrimage to the small archipelago to see not vast numbers of bird species, for the list is small, but rather the albatross, the penguin, the cormorant, three boobies, two frigatebirds, a tropicbird, and two rather odd gulls. And the finches of course, but they are discussed in the next chapter.

The oceans appear vast and uniform to the human eye. We watch the waves. The winds' collective energies can raise whitecaps, create huge swells, or cease altogether, flattening the sea into a state of idle calm. We observe the intense blue coloration, a fact of physics, the scattering of light at blue wavelengths as solar radiation penetrates water. The ocean looks unidimensional. But not to a seabird's eye. To seabirds, oceans are very different from place to place because the waters are profoundly influenced by patterns of currents and currents affect, actually determine, patterns of life within the sea. Some currents are dense with chemical nutrients and thus abound in plankton. The plankton is so rich that the diminutive plants and animals form an immense base of a complex food web that effectively channels solar energy through the diverse food chain all the way to the largest creatures extant on the planet, the great whales. Other currents are, for a variety of reasons, essentially barren of plankton and offer little sustenance to vertebrates of any kind. Birds have adapted not only to fly seemingly effortlessly over the restless waves of the world's oceans but also, and most importantly for them, to find where the food is.

The Galápagos Archipelago is not an oceanic desert, at least not when currents are flowing normally, a pattern that changes during years of El Niño events. On the contrary, the Galápagos offer seabirds excellent fare, an abundance of fish, squid, and other food sources both around the islands themselves and at sea away from the archipelago.

Unlike mammals and possibly reptiles, which evolved life forms that became totally independent of the land, birds must return to land in order to breed. The extinct reptilian plesiosaurs and ichthyosaurs of the Mesozoic Era may or may not have had to go on land to lay eggs. Fossil evidence suggests that the fishlike ichthyosaurs likely gave birth to live young, which would have made these creatures independent of land. All birds, however, lay eggs, and without question, all birds' eggs must be laid terrestrially.

Among the seabirds of the world, it is the seventeen species of penguins that are most adapted to life at sea, indeed, *in* the sea. These birds literally fly through the water, not the atmosphere, using their wings for propulsion as they pursue their prey below the surface of the waves. Penguins are sleek, torpedolike in body form, and marvelously insulated by fat and feathers against the chilling effect of the frigid water in which they move, but not a single penguin can reproduce without returning to land. The most southern bird species in the world, the emperor penguins *(Aptenodytes forsteri)*, huddle together in dense breeding colonies to maintain body warmth during the depths of the dark antarctic winter, the penguins having walked far from the edge of the sea to incubate a single egg tucked between the feet and the body to keep the egg warm. Though the world's fourteen albatross species are champions of long-distance travel, able to traverse thousands of miles of seemingly trackless ocean in search of prey, each and every one of these birds, superbly adapted to fly but ungainly on land, must nonetheless return to land in order to court, mate, nest, and care for its young.

Land is at a premium for seabirds. The long and narrow wings that typify many seabirds, while impressively adapted for dynamic soaring just above the waves, are poorly adapted for land, where air currents behave quite differently. Of course, nesting along coastal areas is possible, but coastal areas offer opportunities for predators. And most seabirds are colonial, so the invasion of a colony by predators can quickly annihilate the colony. Islands offer isolation, a major protection from most predators. Islands like the Galápagos, located in a reasonably rich area of ocean, are ideal.

A memorable line in the wonderful film *Field of Dreams* is "If you build it, he will come." Earth built the Galápagos, and the seabirds came. And they came from all over. Two booby species colonized the islands from the tropical Pacific; one came from the western coast of America. One frigatebird species arrived from the Caribbean; one came from the tropical Pacific. A penguin swam and a pelican flew with the Humboldt Current to the Galápagos, stayed, and evolved. A tropicbird colonized from the Caribbean, while two kinds of terns came from the tropical Pacific. A cormorant came, probably from the western coast of America—no one knows for sure—and evolved into the world's largest of its kind, giving up flight in the process. A gull came, likely from somewhere along the long western coastline of the Americas, and evolved into a new species, while another gull, a different gull, came and evolved into yet a different species with large eyes, very large eyes, but no one is willing to say from where this odd creature emigrated. We just don't know. The enchanted isles retain many mysteries.

Seabirds offer wonderful opportunities for study. Ecologists and evolutionary biologists have devoted much effort to unraveling the life of seabirds because these creatures collectively represent diverse and fascinating solutions to the same problem: how to wrest a living from the ocean. Some seabirds, like the boobies, dive into the sea from above, avian bombs that become feathered torpedoes, their webbed feet speeding them through the depths to grab fish. Others, like penguins, dive from the surface of the water and beat their wings to move while submerged. Albatrosses do not dive at all but usually rest atop the water and snag prey with their long hooked beaks, a behavior called *surface-seizing*. And still others, like the frigatebirds, often rob other seabirds of their catch, harass them in flight until they drop or vomit their prey, and then adeptly catch it midair. The frigatebird is a feathered pirate. Natural selection is not necessarily moral. If it works, it evolves.

Seabird colonies are scattered widely around the world, but the Galápagos Islands offer a collection of these birds that is both unique and easily seen. They host virtually the only breeding colony for one of the world's albatross species and harbor the most northern of the world's penguins. And one remarkable species, the flightless cormorant, has no parallel elsewhere. It is the world's only cormorant that has lost its ability to fly.

In order to see the full array of Galápagos seabirds, you must visit a number of key islands. Bolivar Channel, between Fernandina and Isabela, is essential, for it is there that you can regularly find the flightless cormorant. Bolivar Channel is also a good place to see the Galápagos penguin, though this species can be seen well elsewhere, especially around Bartolomé. Española is one of only two islands where the waved albatross breeds (though the bird can sometimes be seen at sea around the islands), and Genovesa is where to go to see red-footed boobies and wedge-rumped storm petrels at their breeding colonies. Other seabird species may be encountered fairly widely within the archipelago, such as the blue-footed and masked boobies, Audubon's shearwater, the dark-rumped petrel, Elliot's storm petrel, the two frigatebird species, the red-billed tropicbird, and the gulls.

This chapter provides an account of the resident seabird species, but you should keep in mind that birds have wings and tend to use them. In a recent field guide to birds of the Galápagos, there are just over three-dozen species of water birds—storm petrels, albatrosses, gulls and terns, and shorebirds—listed as "vagrants," so it is well worth keeping a sharp eye and not assuming the odd bird to be a routine sighting. It may not be.

In Darwin's account of his days on the Galápagos, there is essentially no mention of seabirds. For the most part these species would probably not have impressed Darwin because, by that time in his five-year journey, he would be used to seeing boobies, storm petrels, gulls, and others at sea and coastally. Unfortunately, when the HMS *Beagle* sailed through Bolivar Channel, and when Charles Darwin briefly visited Isabela (Albemarle) at Tagus Cove, he apparently did not see either the penguins or the flightless cormorants. Rest assured, if he had, he would have written about them.

The Galápagos Penguin

Two penguin species inhabit the southern coasts of South America, the jackass penguin *(Spheniscus demersus)* and the Magellanic penguin *(S. magellanicus)*. The latter is named for Magellan, the explorer, and the former for its peculiar, donkeylike braying. Both are small penguins, about 27 inches tall (70 cm), with a black stripe (vaguely suggesting a kind of "vest") separating their black backs from their white bellies and varying amounts of pink around their eyes and inner beaks.

A third species, the Humboldt penguin *(S. humboldti),* bears a close resemblance to the other two but is more northern, inhabiting offshore islands of Chile and southern Peru. The cold Humboldt Current runs along the South American coast, and occasionally a Humboldt penguin swims as far north as Ecuador, virtually on the Equator. Recall that the Humboldt Current swings westward as it approaches the Equator. Recall also that it remains cool, even when reaching the Galápagos. Finally, recall the Cromwell Current is a slow, deep, cold current that flows east and upwells at the western part of the islands when it runs up against the Galápagos plateau. The picture becomes pretty obvious. Carried by the westward-flowing Humboldt Current, the ancestors of today's Galápagos penguins found suitable habitat abundant with food throughout the westernmost parts of the archipelago, where upwelling keeps the waters at their coldest and most productive. Presumably the Galápagos penguin *(S. mendiculus)* is descended evolutionarily from the Humboldt penguin. They certainly look alike, though the Galápagos species is considerably smaller, at a mere 20 inches (about 50 cm).

Galápagos penguin

Look for Galápagos penguins not only on the rocks bordering Isabela and Fernandina but also around Floreana and particularly around Bartolomé and the channel that separates Bartolomé from Santiago. Like virtually all Galápagos species these animals can be oblivious of humans, even when said humans are in extremely close proximity. Still, keep a respectful distance.

Penguins feed by pursuing prey under water. One of my most memorable natural history experiences was swimming with Galápagos penguins and observing how they went about the job of having lunch. I had seen and photographed penguins earlier in the morning while visiting Bartolomé. The animals were resting on some protruding rocks near Pinnacle Rock, and I literally waded through the shallow water holding my camera with telephoto lens well above my head as I approached relatively close to the birds. As far as I could tell, my presence meant nothing to the two penguins, and the photos are among my favorites of that trip.

Later my group visited the pahoehoe lava field on Santiago, but a friend and I opted to remain at the landing beach and snorkel. The heat at midday was overpowering, and the thought of hiking across a black lava field did not appeal. Once in the water, snorkel and mask in place, my friend and I were a subject of great curiosity to two sea lions. After some "swim by" perusals and at least one gentle, inquisitive nip on one of my flippers, the sea lions went about their business. Perhaps they bore easily.

The water was glassy clear, and there were scattered, dense schools of tiny, silvery, sardinelike fish swimming mostly near the surface. Out of the corner of my eye I saw a shape, and suddenly a small feathered torpedo with wings appeared, quickly followed by another. The two penguins were soon actively feeding on the tiny fish. I knew from the ornithological literature that Galápagos penguins often feed in pairs. The birds would swim under the school, which whirled in ever tightening circles, and soon the penguins would make some seemingly uncommitted approaches, the effect of which would be to scatter the school, which would then quickly regroup. Again the penguins would close in and scatter the fish. As the fish dispersed, the birds would quicken their pace, turning with great speed in pursuit. I saw one penguin literally capture a fish a few feet in front of my facemask. The fish generally kept their distance well, but the penguins were good at what they did too.

I have rarely observed the adaptive advantage of schooling behavior so clearly. Clouds of tiny fish would move to form a kind of halo around a penguin, and the penguin would make its move. What it came down to was this: If all, and I mean all, of the fish sharply turned in the same direction, the penguin would come up empty, with no catch. But should any fish turn differently from the others, the bird would snag the fish—every time, of at least a dozen attempts. It was obvious that the adaptive protection afforded by the school is that the mass of fish effectively prevents a penguin from clearly homing in on one fish and following that particular fish to the point of capture. The constant turmoil of motion in the school allows each and every fish to be lost in the crowd, so to speak, the maelstrom of moving fish becoming an effective shield for each and every member of the school. But woe betide the fish that misses a crucial turn. Instantly that individual is no longer under the protection of the school, and it becomes destined to spend the rest of its day being digested within a penguin.

Galápagos penguins nest in the small caves, burrows, and cracks found in the rocks that line the shores near where the animals feed. They are not tightly colonial, but they nest only around Fernandina and Isabela. Like most bird species they form a pair bond, and both parents incubate the two eggs, a period normally requiring about forty days. In good years when food is abundant, a penguin can produce three clutches. Nesting can occur anytime during the year. What is important is food availability. This species is one of the most negatively affected by El Niño events. From an estimated population of between six thousand and fifteen thousand in 1977, approximately 77 percent of the population was lost during the severe El Niño of 1982–1983. A census conducted in 1984 recorded a meager 463 birds. But the penguins persist, their population slowly increasing, an indication that they can bounce back from such calamities. The most recent population estimate is about twelve hundred birds, which suggests that the population is slowly recovering from the devastation of El Niño. Unfortunately reports suggest that the birds can suffer negative effects by disturbances caused by fishermen and occasionally tourists. Nests sometimes are preyed upon by such introduced mammals as cats, rats, and dogs. Like other Galápagos endemics, these birds need all the help they can get.

The Flightless Cormorant

One of Charles Darwin's positions in the "one long argument" he titled *On the Origin of Species* was that there would be evolutionary consequences as a result of what he called "use and disuse" of various anatomical parts. He wrote of island populations of insects, beetles in particular, and how they would tend to lose their ability to fly. He attributed the loss to natural selection. The argument is simple. A small insect, blown out to sea, would quickly perish. Although it might be advantageous for a continental insect to have the dispersal abilities provided by flight, it could be counterproductive to the reproductive potential of individuals situated on islands. So those individuals least apt to fly, by virtue initially of disuse and later of mutations that reduced wing size, would have a selective advantage over other "normal" flying individuals and would thus leave more offspring. The genetics of wing reduction would be promoted generation by generation, resulting in flightlessness among island beetles. Darwin also noted that several duck species have wings so small as to render them flightless and concluded that "I believe that the nearly wingless condition of several birds, which now inhabit or have lately inhabited several oceanic islands, tenanted by no beast of prey, has been caused by disuse." Oh, would he have liked seeing the Galápagos flightless cormorant *(Phalacrocorax harrisi)*.

The flightless cormorant is the only member of its family, the Phalacrocoracideae, to have lost the ability to fly. It is one of thirty-nine species of cormorants, all of which dive from the surface in pursuit of fish. It is sufficiently unique that some taxonomists place it in a separate genus, *Nannopterum*. The bird is extremely large for a cormorant, with a body length reaching 39 inches (100 cm) and weighing almost 9 pounds (4 kg). By comparison, the Guanay cormorant, an abundant species along western South America, is more typical of cormorant dimensions, with a length of 29 inches (76 cm) and weighing about 5 pounds (2.25 kg). The flightless cormorant is grizzled brown with a long beak, hooked at the tip, large, penetrating blue eyes, and immense feet. Most noticeable are the wings, or rather the almost lack of them.

The wings are disproportionately small, to put it mildly, making the birds almost comical in appearance when they spread their wings, in typical cormorant fashion, to dry them after diving for food. Though the wing area is small, the birds have particularly large feet that propel them in quest of fish and squid.

There are several factors that combine to select for flightlessness, and Darwin certainly identified one of them, and it wasn't merely "disuse." Darwin noted that flight could in fact be a high-risk behavior for island populations well isolated from other lands. Flightless cormorants are extraordinarily sessile, remaining essentially where they were hatched for their lifetimes, which is a selective advantage because there are only a few places on the Galápagos where cold currents supply the food required by these animals. Of course they also evolved in an essentially predator-free environment, removing the necessity of flying to avoid capture. This, of course, is now a potential risk, and efforts must be unceasing to provide adequate protection for these birds. Finally, as mentioned in the discussion of the giant tortoises, large body size often evolves in island populations. As cormorants added weight, flying became more costly (added weight requires disproportionately greater wing area) and, at the same time, less adaptive. The result was that flight abilities atrophied, while swimming and diving abilities, crucial for gathering sufficient food, were enhanced. Indeed, flight has atrophied in the most literal sense: Not only are the wings tiny in relation to the bird but the huge keel, the carina, projecting from the breastbone and anchoring flight muscles on all flying birds has also become vestigial. (Ostriches and their kin lack carinas.)

Flightless cormorant drying its diminutive wings

Flightless cormorants are restricted to the cold waters of the western Galápagos, where the Cromwell Current upwells. Like the penguins, they are hit hard by El Niño events, but they have shown the ability to recoup quickly. When the severe El Niño of 1982–1983 claimed about half of the normal population of between nine hundred and twelve hundred birds, the population nonetheless managed to regain its normal density by the end of 1985.

The breeding biology of the cormorants appears to be successfully adapted to the unpredictable nature of their food source, resulting from the capricious temperature fluctuations of the waters in which they make their living. The birds essentially breed whenever they can, meaning whenever there is sufficient food to carry them through the breeding cycle. In the case of flightless cormorants, being flexible is the key to survival. And being flexible means significant childcare duties for the male. Unlike other cormorant species, female flightless cormorants normally desert the male with whom they bred, somewhere around two to three months after the eggs hatch. Up to that point, the female and male both have fed the young. Now it's the male's job alone, a job that requires an additional commitment of between five and nine months until the young are fully fledged. The female, once she abandons her mate, quickly attempts to breed again, squeezing the maximum amount of reproductive effort out of a season too short for a given pair to fully raise two broods but not too short for a female to cut out on her mate, find another mate, and make a second brood.

Staggering broods, which depends, of course, on the original male remaining to care for the first brood, is a unique solution to the problem posed by the precarious nature of the food source, dependent as it is on constant (but easily interrupted) upwelling. As it is, though three eggs normally are laid, most times only one young cormorant will survive to maturity. Some eggs are sterile, some are preyed upon by Galápagos hawks, and, in most cases, the smaller of the nestlings simply dies of starvation.

Now, about that first male. Why does he remain with the young once his mate has deserted? The answer, as with essentially all of natural selection, is in the genes and what behavioral "strategies" are ultimately good for the propagation of those genes. The young birds are the male's genetic investment as well as the female's. To desert them would be to lose the weeks of effort already invested in the breeding process. But if the male deserted the female, before she deserts him, would not she be left holding the eggs, so to speak, while the male is free to breed again? She would, but males are about 15 percent heavier than females and can dive deeper and remain underwater longer than females. In short, they are better providers because they can catch more and bigger prey. A deserted female would not be able to supply enough food to raise a brood on her own, and such a failure would result in terminating her mate's genetic investment as well as her own. So, in fact, it is not in the male's best genetic interest to desert. He's kind of stuck in the calculus of natural selection.

But there is more. Flightless cormorant society is very inbred. The birds lack any dispersal whatsoever, and so brothers have little choice but to mate with sisters, sons with mothers, and so forth. It's one big family affair. Thus, when a female deserts for another male, it could well be that the other lad is a brother, first cousin, or even the father of her original mate! Suppose it is a brother. If so, it shares 50 percent of its genes by heredity with the first mate. So, if the female breeds successfully with the first mate's brother, each of those young have, by heredity, 25 percent of the original mate's genes. And, if the brother raises one offspring of his own (from the first mating) and the original mate raises one chick, the brother will have 75 percent of his genes duplicated in that breeding season (50 percent from the one chick that he fathered and 25 percent from the chick that his brother, the original mate, fathered). But consider that the female may be related directly to the first male, perhaps his sister. If she deserts him for his brother, and two chicks are raised, it works out that the original male will have seen the duplication of somewhere around 125 percent of his genes by virtue of the interrelationships among the three birds. This is not as far-fetched as it may sound given the extreme inbred nature of the population. But inbreeding has costs, and the fact that eggs are often infertile may be suggestive of inbreeding depression.

The best place to observe the flightless cormorant is Punta Espinosa on Fernandina, where there are normally about one hundred pairs nesting in the immediate vicinity. You may even see the birds doing their courtship behavior, walking in synchrony with exaggerated steps, tails cocked, their necks snaked at a sharp angle. They look stiff and strange, but they apparently find the behavior satisfying. They also court when swimming—passing each other, periodically looking skyward, raising themselves somewhat out of the water, and flapping their diminutive wings. When you observe these curious evolutionary novelties of the Galápagos, think again about how their odd breeding ecology and prolonged breeding season are the products of an unpredictable food environment and are influenced by the inbred nature of the population. If only Darwin had known.

Boobies

Booby. The name itself sounds like the source of a joke. "Have you seen a booby today? Oh yes, lots of boobies. Boobies everywhere." Well, if you travel to the Galápagos, you will encounter boobies. No one seems certain as to why this group of splendidly adapted, graceful seabirds has such a dismissive common name—their temperate kindred are called *gannets,* a perfectly respectable-sounding name. Boobies are very tame, showing essentially no fear of humans, a characteristic that has made a meal out of more than one of them and kept quite a few stranded mariners in calories long enough to be rescued. So, the story goes, they are called *boobies* because they are dumb, never learning to fear the most fearsome species on the planet. Another frequent suggestion is that the name derives from the Spanish word for clown, *bobo,* and that could be true given the comedic nature of booby courtship behavior.

Boobies are seabirds in the family Sulidae (boobies and gannets), a group of only nine species, ranging widely over the world's oceans. Their nearest relatives are such birds as pelicans, cormorants, frigatebirds, and tropicbirds, all of which have species that reside in the Galápagos. So you can venture to the Galápagos and see one-third of the sulids of the world. And rather easily at that.

There are three species of boobies on the Galápagos Islands. The most frequently sighted on the Galápagos is the blue-footed booby *(Sula nebouxii).* It ranges from the Baja Peninsula in northwestern Mexico to northern Peru, including, of course, the Galápagos Islands, where there is a breeding population of about ten thousand pairs. The second most commonly observed booby on the archipelago is the Nazca booby (*S. granti*), confined to the Galápagos and recently separated from the similar masked booby (*S. dactylatra*), a wide-ranging species found throughout the equatorial oceans of the world, particularly the tropical Pacific. There are estimated to be about twenty-five thousand to fifty thousand breeding pairs of these birds in the Galápagos. The third Galápagos species is the red-footed booby (*S. sula*), which, like the masked booby, ranges very widely over tropical oceans, including the Caribbean, where there is a large population. There are estimated to be about 250,000 pairs of red-footed boobies on the Galápagos. However, the red-footed booby is the least frequently sighted of the Galápagos boobies.

Oddly, the least abundant Galápagos booby, the blue-footed, is the most often encountered. This is because blue-footed boobies are inshore feeders and are often seen plunge diving in the bays throughout the archipelago. Masked boobies feed offshore but not generally as far offshore as the red-footed. The red-footed booby is the least frequently sighted of the three for two reasons. One is that 140,000 pairs of the red-footed boobies reside on Genovesa, and not all tour boats visit Genovesa, which is rather far to the north of the main islands. No birder should ever select a trip to the Galápagos that fails to go to Genovesa. It is truly a spectacle for those who enjoy feathered creatures. And you might not see red-footed boobies if you don't go there. The other reason is that these birds fly far out to sea to feed, dispersing widely, and thus may not be seen around the islands themselves.

Boobies are graceful birds with long wings, wedge-shaped tails, and spearlike bills. They fly purposefully, occasionally gliding, usually in loose groups. Most Galápagos visitors witness at least one "feeding frenzy" of blue-footed boobies. When boobies are plunge diving, it is particularly wonderful to be snorkeling and be able to watch the birds below the waterline. When a school of fish is concentrated and the boobies discover it, they gather above and begin plunge diving, folding their wings back and plunging headfirst into the sea. These dives, often executed from considerable heights, can be quite spectacular. The masked booby has been observed diving from heights of 328 feet (100 m), though heights of between approximately 30 and 100 feet (10 and 30 m) are more common for most species. The blue-footed booby is a shallow water diver, its tail somewhat longer than that of the other booby species, allowing maximum maneuvering once below the water.

Boobies have excellent binocular vision. Look one straight in the eyes, and both of its eyes will look straight back at you. Effective stereoscopic vision permits the birds to identify prey with great precision before beginning the dive. It would be interesting to learn if these birds have the ability to reduce glare on water, such as "polarized vision," for example. Booby eyes are large, allowing for good underwater vision in a light-reduced environment. Some are capable of diving to depths of nearly 80 feet (25 m).

Once below the surface, boobies often beat their wings to gain added speed as they pursue fish. Some boobies will occasionally dive from the surface rather than plunge dive, and red-footed boobies are known to effectively swoop down and capture flying fish "on the wing."

The three Galápagos booby species are each easy to identify. Blue-footed boobies are brownish above, scaly on the back, and white below. Their necks and heads are streaked with brown. And they have large blue webbed feet—very blue feet. Red-footed boobies come in two color phases. Most common on the Galápagos is the brown morph, essentially uniformly rich brown. The white morph is white with black wing feathering and a pinkish tone to the neck and head. And both morphs have large red webbed feet—very red feet. The Nazca booby is named for the dark black surrounding its face and eyes and is basically white with black wing feathering, so it can, in flight, occasionally be confused with the white morph of the red-footed booby. The large webbed feet of a masked booby are basic black.

All boobies are colonial. There are many small blue-footed booby colonies scattered throughout the islands, and observing blue-footed booby courtship behavior is one of the high points of most Galápagos trips. Masked booby colonies are somewhat fewer but occur on enough islands that most visitors see this species with no difficulty. Red-footed booby colonies are situated only on the outlying islands, the largest on Genovesa, which is visited by many, but not all tours. Both blue-footed and masked boobies nest on the ground, while red-footed nest in shrubs and small trees. In the case of the first two species, no actual nest is constructed; only a bare scrape is used in which to lay the eggs. Eventually, excrement from the parent birds defines a "nest area." Red-footed boobies make small nests of twigs.

Courtship is a big deal in boobyland. Sulids must establish firm pair bonds because both birds are essential in raising the young. To seal the pair bond, there is ritualized courtship behavior. Since it is the blue-footed booby whose sex life is most often observed by Galápagos tourists, I describe its courtship here. The others' courtship behavior is basically variations on this theme.

First, you must learn to tell the male from the female. It turns out that's pretty easy because you can normally come within about 5 feet (1.5 m) or less of the boobies you want to observe. Female boobies are just a tad larger than males and have what appear to be big black pupils, giving their eyes a "dilated look." Males, smaller, have little pinpoint pupils. Actually, so do the females, but females also have black pigmentation around their pupils that effectively makes them seem much larger. Boobies vocalize with great regularity, and females and males sound quite different from each other. Females honk, sounding like a goose, while males make a high, shrill, whistling sound that always makes me think they have a nasal obstruction.

Once the pair is in proximity, the male may begin by opening its wings at an odd angle, cocking its tail, and looking directly upward, bill pointing to the zenith, the bird apparently mesmerized by its view of the universe. This behavior is termed *sky-pointing*. Other behavior follows, often involving the male walking with exaggerated steps around the female, his tail cocked upward. The blue feet are indeed emphasized throughout the various behavioral patterns, and it is likely that the intense blue coloration evolved in response to selection pressures for species recognition (boobies nest in mixed species colonies, where confusion among species would elicit a strong negative selection pressure) and for courtship bonding such that each bird's reproductive cycle is stimulated and coordinated.

On occasion a male may present a female with a stick, a seemingly odd behavior in that they make no nest. Perhaps ancestral boobies constructed actual nests, and thus nest building is the "default position" in the genes of boobies. Dropping eggs on bare ground may be evolutionarily more recent, an adaptation to islands where predators are few and sticks even fewer. But it is also possible that the stick serves the function of helping the birds decide exactly where to place their nest. Red-footed boobies do still make nests, and sticks are a big deal to them, providing they are on an island that has sticks. There are sticks on Genovesa, where the red-footed nest, and it is fascinating to observe how frigatebirds, who also nest on Genovesa and need sticks, liberally pilfer booby nest sticks.

Boobies have small clutches, up to three eggs for the blue-footed, only two for the masked, a mere one for the red-footed. In the case of the blue-footed, they will raise up to three chicks if they can, but they often can't. There just is not enough food to be had. This is where asynchronous hatching enters the evolutionary equation. Boobies begin incubating when the first egg is laid, and the eggs are laid about a day apart. That means that the eggs hatch about a day or so apart, which means that the infant birds will not be the same size. The first hatched will be fed immediately, will begin to grow, and will be larger than the second. And the second will be larger than the third. In the case of the blue-footed booby, if there is not enough food, the parents just feed the largest, and the others starves. If there is a booby heaven, the expression "Mom always liked my brother better than me" is often uttered.

As for Nazca boobies, asynchronous hatching results in a slightly different pattern but with essentially the same outcome. Two eggs are laid asynchronously, one hatches first, and the bigger of the siblings simply kills the smaller of the pair, usually by literally evicting it from the nest area. Always. Nazca boobies lay two eggs and always raise but one young. Why? The "insurance policy" hypothesis suggests that the second egg provides an advantage: If the first egg is destroyed, the second

Male blue-footed booby displaying before a female

Blue-footed booby sky-pointing

egg will hatch a chick to be raised. On the other hand the second egg may represent a relic of evolution, when this species was able to raise more than one young at a time. In any case, like much of evolutionary biology, it's pretty weird.

Frigatebirds

A seabird that can't swim? That cannot land on the ocean or it will drown? There is such a creature, five species of them, in fact, and two of those species reside on the Galápagos. These are the birds in the family Fregatidae, the frigatebirds. Described as "pantropical," each of the five species roams some part of the world's tropical seas. One highly restricted species lives only on and around tiny Ascension Island in the South Atlantic. But the two frigatebird species found on the Galápagos are much more widespread. The great frigatebird *(Fregata minor)* ranges widely throughout the Pacific and Indian oceans and undoubtedly colonized the Galápagos from the west. The Galápagos represent fundamentally the easternmost range of this species. The magnificent frigatebird *(F. magnificens)* is distributed coastally, along both the western and eastern shores of the Americas. There is also a small population that colonized the Cape Verde Islands, off the west coast of Africa. This species undoubtedly colonized the Galápagos from the east, and the archipelago represents essentially its westernmost range.

It's true: Frigatebirds cannot swim. If they get wet, they drown. Birds wet from rain can remain in flight until they reach land, but should the birds put down on the water, their feathers will become too saturated for them to take off again, and that's pretty much that. The reason is that these birds have a tiny uropygial gland. Located just at the base of the tail, the gland normally supplies birds with oil that they smear on their feathers to render them waterproof. So, not surprisingly, there is a very strong selection pressure constantly asserted on frigatebirds to remain above the waves, even as they snatch prey from the surface, using their long hooked beaks. Frigatebirds are among the premier aerialists of the bird world.

Both frigatebirds of the Galápagos have similar shapes. They look like giant blackish swallows, with long and slender pointed wings and scissorlike tails that allow them precise control of their flight. The wingspan approximates 7 feet (2 m), so these are rather imposing birds. They are well adapted for both gliding and pursuit flight. One of the common sights around the islands is masses of these birds flying effortlessly above, like so many skinny kites. But equally common is the sight of a frigatebird in swift pursuit of a booby or gull and forcing it to disgorge its recently captured food, the frigatebird adeptly snatching the falling item before it reaches the ground. This kleptoparasitic behavior earns the birds their alternate name, *man-o-war birds*. Frigatebirds are capable of capturing food on their own and forage often far at sea, where they prey upon surface-dwelling fish, particularly flying fish and squid. Their long bills, sharply hooked at the tips, plus their superb flight skills permit them to capture prey, usually without hitting the water itself.

There are considerably more pairs of great frigatebirds on the Galápagos than there are of the magnificent frigatebirds. One of the largest colonies is on Genovesa,

Male great frigatebird, wings spread

where twenty thousand pairs breed. In contrast, there are estimated to be only about one thousand pairs of the magnificent throughout the archipelago. Magnificent frigatebirds can be easily seen at their colony on North Seymour, where they commingle with the great within a single colony.

The males of both frigatebirds, being almost uniformly blackish, look alike in the air and are thus hard to identify. Females are much easier. The female great frigatebird has white on the throat extending to the belly, while the female magnificent has a brown throat and a white band on the belly. But most visitors to the islands pay relatively little attention to identifications of these species in flight because they can walk right up to the birds as they reside on their nests, and then the identification becomes rather easy. Male great frigatebirds have a decidedly green sheen on their back feathers, while male magnificent have a purple sheen. Birders use the expression "Green is great!" to associate the green coloration with the great frigatebird. Females, as noted above, are easier to tell apart, but at close range you can see the decidedly red eye ring of the great frigatebird female, which contrasts with the blue eye ring of the magnificent female.

The most impressive characteristic of frigatebirds is arguably the throat sacs, which ornithologists call *gular sacs,* of the breeding males. Males of both species have extensible bright red skin on their throats, hardly noticeable except when filled with air. Then the sacs swell like balloons, dangling beneath the necks of the birds, an advertisement of male fitness, all compliments of natural selection. Males inflate their throats both at rest and in flight. Displaying males sit conspicuously on a small

Great frigatebird pair, male with throat pouch deflated

tree or bush, fully inflate their throat sacs, spread and shake their wings, look skyward, and vocalize. The two species also sound quite distinct. The male magnificent makes a flat, clattering sound, but the male great, my favorite, emits a kind of high-pitched, oscillating whinny, sounding like flying saucers landing in a scene out of an old 1950s sci-fi movie.

Most ecotourists get to see frigatebird chicks in nests. Only a single egg is laid per nest, and both parents incubate. Chicks hatch naked but are soon covered with fluffy white down. They remain nestbound for five to six months before attaining sufficient plumage to fledge. Parents attend the chicks throughout this period. The nest itself is but a small platform of sticks that does not inspire great confidence in the birds' architectural skills. Sometimes the egg or newly hatched chick falls from the precarious nest and perishes. Chicks who fall are also prey to hawks and other frigatebirds.

The Red-Billed Tropicbird

Like their close relatives the frigatebirds, tropicbirds, family Phaethontidae, are pantropical, the three species widely distributed over equatorial seas. It is the red-billed tropicbird *(Phaethon aethereus)* that calls the Galápagos Islands its home, or at least one of them. This species actually has the most restricted range of any of the three, though it can be found in the Caribbean, South Atlantic, and Indian oceans. It is fundamentally absent from the Pacific, save for the population of several thousand pairs on the Galápagos.

Tropicbirds are, in a word, gorgeous, and these birds seem to epitomize tranquil tropical oceans. About the size of a pigeon, they resemble terns that got really dressed up. Each of the three species is white with variable amounts of black highlighting the plumage, and each has a long, streamerlike tail that trails gracefully behind the bird in flight. The red-billed tropicbird has black outer wing feathers and wavy black lines on its back, extending to the upper base of the tail. It has a defining black stripe through its eyes and, as the name implies, a blood red bill.

Tropicbirds are widespread throughout the Galápagos and are to be looked for anywhere that there are cliffs, for that is where they nest. They construct no actual nest but select a cavelike niche among the rocks as the site where the egg, just one egg, will be laid. The very definition of grace in flight, a tropicbird can look mighty cramped when squeezed among the rocks incubating the egg.

Tropicbirds feed on a diet of mostly fish, with occasional squid. They mostly feed by plunge diving, somewhat like boobies, but they are also sufficiently skilled to take a flying fish in flight.

What is most fun to watch with tropicbirds is their often dramatic courtship flights, when several individuals fly close together, screaming at the tops of their avian lungs, performing twists and turns, and then executing long downward dives where two birds often appear to actually touch wings as one descends upon the other, an aerial performance the likes of which few other species could accomplish. When tropicbirds are performing, they tend to stop the show. Naturalist guides cease talking about lava, or finches, or palo santo trees, and everybody just watches, except, of course, for those who scramble to adjust their camcorders. By the time the performance is complete, it makes you want to applaud. It's that good.

Red-billed tropicbird

The Brown Pelican

Most people know perfectly well what a pelican looks like, a big bird with an even bigger bill, sort of like a mail pouch for fish. Pelicans, in the strict sense, are not seabirds, at least not pelagic ones. They are more appropriately described as coastal seabirds, and some of the world's seven species range well inland, on continents from Australia to North America. The brown pelican *(Pelecanus occidentalis)* ranges from California south to Tierra del Fuego, as well as throughout the Caribbean and Gulf of Mexico. The largest numbers of brown pelicans are found on the guano islands off the coast of Peru, where nearly a million individuals occur. It is clear that sometime in the past a few of these Peruvian birds must have continued cruising the Humboldt Current until they found themselves new residents of the Galápagos. There are only a few thousand pairs on the Galápagos, but the population is recognized as a separate subspecies with the unattractive name of *P. occidentalis urinator.*

Brown pelicans never venture inland but are strictly marine, mostly inhabiting bays and estuaries and rarely venturing far over ocean waves. Unlike any of the other pelican species, they feed by plunge diving, sweeping upward to begin a parabolic dive that ends as they throw their wings back and gracefully plummet into the water. Once in the water the open bill is used to capture any number of small, sardinelike fish. The lower mandible has a huge skin pouch that distends to accommodate a load of fish. This, of course, is the most distinctive characteristic of the group.

The brown pelican is unusual not only for its feeding behavior but also for its plumage because it is the only pelican to be uniformly brown rather than mostly white. When breeding, the adults develop a bright yellowish-white head and chestnut neck, and the pouch becomes colored in various shades of red. Most Galápagos brown pelicans nest among the mangroves in small colonies.

The Waved Albatross

Some birds just have really inappropriate common names. The ring-necked duck (*Aythya collaris*), which occurs throughout much of North America, is certainly one of them. The so-called ring around the neck is visible only when the bird is literally in the hand, and even then you have to look hard. Only a museum taxonomist could have named it. No field birder would have named it that, though the name *ring-billed duck* would be fitting because, even from a distance, this species has a very well marked white ring around its bill.

Same story, with a twist, for the albatross that breeds on the Galápagos. Its official common name is *waved albatross* (*Phoebastria* [formerly *Diomedea*] *irrorata*), a name that derives from the delicate wavy feathering pattern that adorns the breast, upper tail, and mantle (back) of this species. From a boat at sea, you can only imagine the waviness as the bird flies swiftly by while the boat oscillates in the swells. It is not what we birder types call a "field" characteristic. But visitors to Española, where this bird nests, can see the waviness rather well, given that the hefty bird is about a yard away. Consequently, most birders who have seen the waved albatross up close come away quite satisfied with its common name.

Most of the world's albatross species range widely, but the waved albatross is an exception. It is encountered only in the offshore waters off western South America, from northern Chile to northern Ecuador and ranging westward to just beyond the Galápagos, a range encompassing a mere 16° of latitude. There are only two colonies of these birds worldwide, one on Española, where about twelve thousand pairs are known to breed, and another on the tiny offshore Ecuadorian island of La Plata, where a meager ten to fifteen pairs are found. The total population on the Galápagos is thought to be around fifty thousand or perhaps up to seventy thousand birds. Obviously the vast majority of members of this species are dependent on Española, and it is fortunate that this island is one of the most protected among the archipelago.

All albatross species are truly pelagic, rarely venturing close to shore, except, of course, to go to their breeding colonies (which are mostly on islands at sea). They are virtually effortless fliers. When I sailed from Guayaquil to Galápagos, I saw no waved albatross until I was well out to sea, and then the occasional one would fly by, gliding at low altitude just above the waves. Rarely flapping its wings, it relied on dynamic soaring, a process by which a bird utilizes wind energy rather than the kinetic energy of flapping. If the bird is flying with the wind at its back, it pursues a straight course, but if it is flying at angles to the wind or against a headwind, it courses back and forth, changing its altitude and flight angle constantly as it progresses.

Winds are normally quite prevalent at sea. Indeed, they served as the source of energy for virtually all of the great human explorations of past centuries. As wind passes over the surface water of the ocean, it encounters frictional drag. Winds closest to the surface encounter the most friction, so they have the lowest velocity.

Layers of air pass over the waves at increasingly higher velocities as height above the waves increases. Seabirds such as albatrosses have long, slender wings because the leading edge of the wing is the most critical for producing lift—the more edge, the more lift. The birds lose and gain altitude as they fly at varying heights above the waves, slicing upward into speedier air masses, dropping down into slower air, and then flying upward again and gaining ground speed with altitude. Sometimes the long wings appear to actually slice through the waves. It's an amazing way to move. It would be nice to believe they have the intellect to enjoy it.

In aerodynamic terms, seabirds such as albatrosses, shearwaters, and boobies have high "aspect ratios," which is the wingspan squared divided by the wing area. In an albatross the aspect ratio is 15, basically the highest of any kind of bird (by comparison, the aspect ratio of a house sparrow, *Passer domesticus,* is 5.5). Albatrosses also have among the highest "wing loading," which refers to the weight supported by a unit of wing area. Albatrosses are heavy birds, some weighing in excess of 17 pounds (approximately 8 kg). A wandering albatross *(Diomedea exulans),* one of the largest of the albatross species, weighs 6.2 pounds (8.7 kg) and has a wing loading of 140. Only some ducks and swans exceed that wing loading. With such efficient wings, albatrosses can move at speeds in excess of 31 miles per hour (50 km/hour), which enables them to cover a lot of ocean.

The waved albatross, like other albatrosses, consumes fish and squid, sometimes crustaceans, captured at the water's surface. Many of these prey items engage in what is called *diurnal migration,* remaining at the depths during the day and rising to the surface at night. Light attenuates rapidly in water, so diurnal migration is an efficient way for many animals to remain in darkness most of the time, thus reducing the possibility of falling prey to something. Diurnal migration is really a characteristic of zooplankton, and the larger animals, the fish and squid, follow their food sources. In effect, the oceanic food chain literally moves up and down in the course of a twenty-four–hour period. As a result of diurnal migration, albatrosses tend to feed mostly at night, when their prey is near the surface. The birds rest on the surface and snatch food items with their long, slender, hooked beaks. Albatrosses and their relatives, shearwaters and petrels, are apparently among the few birds with an acute olfactory sense; they are able to detect prey by odor as well as vision. Around the Galápagos, some waved albatrosses have been observed attacking boobies and forcing them to dislodge the fish they have captured, which the albatrosses then claim as their own. Biologists call this behavior *kleptoparasitism.* Most other people call it robbery.

If you want to see the waved albatross on Española, make sure you visit the islands between late March and early December. The birds are basically at sea, away from the colony, from mid-December to mid-March, though the odd one might be found. If you go in late March/early April, you will have the opportunity to watch the very curious mating ritual between the sexes. Males typically return to the colony first and stake out small territories in which to court females. The courtship behavior between male and female is elaborate, often termed a dance, as the birds face off, bow to one another, clatter their long beaks together (resembling avian fencers), and alternately gaze skyward, neck stretched and head at a right angle to the body. Animal behaviorists have "official" terms for these various ritualistic actions: *bill-circling, sky-pointing,* and, in the case of the waved albatross, *head-swaying walk,* a

Waved albatross in flight, showing narrow wings

wonderfully comical behavior. The waved albatross swings its head back and forth as it walks, in a most exaggerated fashion, resembling a big feathered metronome.

The nest is really no nest at all but a mere scrape on the ground where the female lays a single egg, usually incubated between the parent bird's large webbed feet. The chick hatches in about two months and does not fledge (fly away) until late December or early January. It might seem problematic that only a single egg is laid, especially since it takes five to six years for a waved albatross to grow to full adulthood. However, adult mortality has been measured at a relatively low 4 percent annually, and an adult can live for upwards of forty years. That provides ample time for an albatross to replace itself in the gene pool. Pairs are believed to mate for life.

Shearwaters, Gadfly Petrels, and Storm Petrels

Albatrosses are members of an order of birds called the Procellariiformes, or *tube-noses,* a term that refers to the elongated nostrils atop the bill. The Procellariiformes also includes such birds as shearwaters, gadfly petrels, and storm petrels. All tube-noses are pelagic seabirds that range extensively over the oceans of the world, unlike gulls, terns, and pelicans, which remain coastal. Like albatrosses, other tube-noses seem to have a sharp olfactory sense that aids in locating prey on the water. Still another essential adaptation shared by all members of the order is the presence of glands that excrete excess salt through the tubed nostrils, permitting the birds to drink seawater and remain in osmotic balance.

Waved albatross. Note tubed nostril, characteristic of Procellariiformes.

The Galápagos provide nesting sites for tube-noses other than the waved albatross. The dark-rumped petrel *(Pterodroma phaeopygia),* sometimes called the Hawaiian petrel, is one of thirty-one species of *Pterodroma* petrels, a group that collectively makes up the gadfly petrels. Of these, only the dark-rumped nests on the Galápagos, though a few other species are occasionally recorded as vagrants around the islands. Dark-rumped petrels are shaped in such a way as to suggest a small and chunky albatross, blackish-brown above and white below, with white on the forehead. Birders look for a characteristic black line on the pale underwings extending from the "armpit" to the front edge of the wing, a field mark relatively easy to see as the bird soars past.

The alternate name, *Hawaiian petrel,* refers to a second population that nests on the Hawaiian Archipelago, at Maui. The species is considered to be at some risk. For example, the Galápagos population, numbering perhaps one thousand pairs, has declined markedly in recent years, a result of nest depredations by dogs and rats. The Hawaiian population was nearly annihilated by introduced mongooses and remains in jeopardy. The breeding success of the Galápagos birds, which nest in burrows in colonies on the highlands of Santa Cruz, Floreana, Santiago, and San Cristóbal, is estimated to be no greater than 20 percent. Visitors are not permitted at the colonies, but the birds are often seen at sea as boats traverse the archipelago. Currently efforts are underway on the Galápagos to monitor the populations, eradicate predators, and even relocate some of the colonies to safer, predator-free islands.

As of this writing, it is anticipated that the American Ornithologists Union will formally recognize the Hawaiian petrel and the Galápagos petrel as separate species sometime during 2002. It is expected that the Galápagos petrel will become *P. phaeopygia* and the Hawaiian petrel *P. sandwichensis*. The name *dark-rumped petrel* will no longer be used.

Similar in appearance to the dark-rumped petrel, the Audubon's shearwater *(Puffinus iherminieri)* can be found along cliff faces and in the seas virtually throughout the archipelago. The species ranges widely over the tropical oceans of the world, but the Galápagos population is recognized as a separate subspecies *(Puffinus iherminieri subalaris)*. Audubon's shearwater is a smaller bird (12 inches, or 30 cm, in length) than the dark-rumped petrel (17 inches, or 43 cm) and shows more contrast, neatly dark above, white below, with no underwing lines of any sort. The name *shearwater* is suggestive, a clear reference to the dynamic soaring that characterizes most of the group, the wings appearing to occasionally "shear" through the waves. These birds can sometimes be observed exhibiting their range of feeding behavior, which includes plunge diving for fish and squid, diving from the water's surface, or seizing prey directly at the surface.

The other group of Procellariiformes to nest on the Galápagos is the storm petrels, of which twenty species roam the world's seas and three are resident on the Galápagos. Sometimes called *sea swallows,* they are smallish birds, the larger ones being only about 8 inches (20 cm) in length. Most are dark blackish-brown, and some species have white below or on the rump. They are not dynamic soarers but have shorter, wider wings than albatrosses and shearwaters and flap restlessly over the waves. Flapping is of major importance in food acquisition because these birds typically hover over the water, their legs extending down to the water surface, their feet pattering delicately over the waves. They typically nest in large colonies where pairs make burrows in which to locate the nests. Storm petrels generally feed at sea during the daylight hours and return to their breeding colonies under cover of darkness, a behavior believed to be an adaptive response to predation by diurnal birds such as hawks, crows, and gulls. There is an interesting twist to this pattern on the Galápagos, however, as you shall learn below.

The south end of Genovesa is home to a vast colony of storm petrels. Two species, Madeiran *(Oceanodroma castro),* formerly called band-rumped, and wedge-rumped *(O. tethys),* nest there, with the wedge-rumped the more numerous. Storm petrels emit a highly musky odor, probably a means by which each bird can help relocate its own burrow, given the density of the colony. Visitors cannot enter the storm petrel colony but can actually smell the birds from some distance away, so dense are their numbers. And birds are active at the colony all day and all night.

To see the storm petrel colony, you must cross Darwin Bay by panga to a cliffside, actually a caldera wall, to a place called Prince Phillip's Steps. The panga ride offers fine views of nesting red-billed tropicbirds and the awkward cliffside landings that these otherwise graceful fliers are forced to make because of their very short legs. Lower on the cliffs, at the water's edge, the grizzled faces of Galápagos fur sea lions gaze out from the shady protection of the rocks. Once on the top of the caldera, there is a short walk through a forest of palo santo trees heavily tenanted by red-footed and masked boobies. The short trail winds to a flattened volcanic ridge where the labyrinthine rocks make suitable nest sites for hundreds of storm petrels, mostly wedge-rumped. The vista is amazing, with masses of storm petrels coursing about over the colony in swirling, restless, seemingly endless flight, a universe of birds, and smelly ones at that.

Another, larger bird flies among the storm petrel hordes, this one mothlike, its wide buffy wings flapping in and out among the throngs of dark storm petrels: a short-eared owl *(Asio flammeus)*. Unlike most owl species, short-eared owls hunt by day and by night, a pattern that may account for why the storm petrels seem to show no distinct pattern in their comings and goings. Owls are a threat to them anytime. Indeed, there is ample evidence of storm petrel predation. Wings, feet, and partially eaten carcasses are scattered here and there. Short-eared owls are found extensively in the Americas, Eurasia, and parts of Africa. The population on the Galápagos is recognized as a separate subspecies *(A. flammeus galapagoensis)*, noticeably darker, especially on the face, than any other subspecies. And, interestingly enough, the owl's usual prey, the wedge-rumped storm petrel, is also designated a separate subspecies, which, confusingly, is *tethys,* the same as its specific name, making the creature officially *O. tethys tethys.*

One other storm petrel, Elliot's *(Oceanites gracilis),* is actually the most frequently seen storm petrel around the islands. The other two species are rarely away from their breeding colonies, opting to fly well out to sea to feed, but Elliot's storm petrel is common in bays and near shore. If you see a delicate black bird with a small white rump skittering around your boat, especially after some scraps have been tossed overboard, it's an Elliot's. When I last visited the islands, I watched several Elliot's storm petrels dancing on the water as they consumed fish oil that was being dumped from the boat's freezer compartment (a wooden box with melted ice in it) after the crew removed what appeared to be a semiputrid (or semifrozen, depending upon how you choose to look at it) pile of piscines presumably once destined for our dining table. The Galápagos population of Elliot's storm petrel is given separate subspecies status *(O. gracilis galapagoensis).* What is most unusual about the Elliot's is that no one seems to know where it nests, though it is strongly suspected to be a resident breeder. Colonies are known to be in Chile and are suspected to be in Peru as well, but the breeding grounds of this species remain a significant ornithological mystery.

The Lava Gull

While watching Elliot's storm petrels feeding on discarded fish oil, I noticed yet another bird, one that was far more interested in the fish than the oil. A sooty-colored gull that had been perched on the bridge of the boat flew down and stood, expectantly, on the rail. Charcoal gray in body with a black head and eyes outlined in white, this was the lava gull *(Larus fuliginosus),* one of two endemic gull species on the Galápagos and the rarest of the world's fifty-one gull species. Only about four hundred pairs of lava gulls exist, all inhabiting various places among the Galápagos. The well-named gull can be encountered pretty much anywhere around the archipelago, usually on beaches or docksides, and the species is particularly common around Academy Bay on Santa Cruz. It is not a colonial species, opting to defend a large territory within which the nest is located. It was not until 1960 that a nest of this species was discovered and, by 1970, only six nests had been found, mostly in proximity to saline lagoons.

Lava gulls are solitary opportunists, feasting on such delicacies as sea lion afterbirths and fish parts discarded from boats. They are also predatory, capturing and consuming baby iguanas and small fish, as well as occasional nestling birds or birds' eggs. On occasion a lava gull will stalk and harass a lava heron *(Butorides striatus sundevalli)*, kleptoparasitizing its catch, which is usually a crab.

Evolutionary biologists are confident that they know the ancestral species that gave rise to the lava gull, something that is by no means clear for some other Galápagos endemics. There is a widely ranging species, common in winter all along the west coast of the Americas, called the laughing gull *(L. atricilla)*. A lava gull looks like a laughing gull that was used to clean a chimney. Lava gulls are ever so slightly larger than laughing gulls but are otherwise anatomically quite similar. In breeding plumage, laughing gulls have dark heads and, during all seasons, dark wings and mantle. It would not take much of a genetic change to increase the melanistic nature of a laughing gull, and that seems to be what produced the lava gull. It is hard to know just why melanism is such a prominent characteristic in this species. Unlike marine iguanas, lava gulls are homiothermic and do not "need" to soak up the sun's energy to warm up. They seem not to be in a predator-threatening environment, so the fact that they are fairly cryptic among lava rocks and beaches may be inconsequential to their survival. It has, however, been suggested that dark plumage helps prevent frigatebirds from seeing these gulls and robbing them. But that would suggest a very strong selection pressure exerted by frigatebirds upon the gulls. Another adaptive model suggests that dark plumage birds in windy environments stay cooler than light-colored birds.

The Swallow-Tailed Gull

The sooty black of the lava gull makes it both distinctive and distinguished, but the other gull with which it shares the islands is perhaps even more attractive. This is the swallow-tailed gull *(Creagrus furcatus)*, a gull so unique that it earns placement in its own genus. The name refers to the forklike shape of the tail, though the tail is not nearly so deeply "swallow-tailed" as is that of a frigatebird or a barn swallow. It is quite a magnificent-looking gull, adorned with a black hood, grayish upper breast, gray mantle, and black wingtips. Most prominent are the bird's very large eyes, surrounded by bright red eye rings. The bill is black with a contrasting white tip.

Swallow-tailed gulls are common throughout the archipelago, though they are least encountered in the western colder waters. The unique characteristic of this species is that it is specialized to feed at night, on squid. Recall that squid are among the animals that rise to the surface at night to feed on plankton. The gull's unusually large eyes certainly must be an evolutionary response to this behavior. Frigatebird piracy has been frequently suggested as contributing to the evolution of the nocturnal foraging behavior of the swallow-tailed gull. Like most gulls, however, swallow-tailed gulls are opportunistic and take fish and other prey during daylight hours.

The birds breed in loose colonies on cliffs and ledges, sometimes on flat areas,

Swallow-tailed gull

often in places where ecotourists can easily see the nests, making for excellent photo opportunities. One study showed that nesting birds have a strong tendency to face the cliff while sitting on the nest. Cliff-facing behavior is commonplace in such gulls as kittiwakes *(Rissa tridactyla)*, a species that nests exclusively on cliffs. The utility of cliff-facing, its adaptive value, is thought to be that the wings and legs of the incubating bird are positioned to prevent eggs or chicks from falling off the precipice. There is no close genetic relationship between swallow-tailed gulls and kittiwakes, so it is believed that the selection pressure to evolve cliff-facing operated independently on swallow-tailed gulls, and continues to do so.

Which brings up the question of just what was the ancestral species? Swallow-tailed gulls range widely when not breeding, but only one small breeding colony exists outside of the Galápagos, on Malpelo Island, off western Colombia. Some other species must have colonized the Galápagos and given rise to the swallow-tailed species. But which one? No one is sure. Those willing to take a position suggest Sabine's gull *(Xema sabini)*, a species that nests circumpolarly on the arctic tundra but migrates south to winter along West Africa and western South America. The overall patterning of the two species is suggestively similar, and Sabine's gull also has a black bill tipped with yellow (the bill of the swallow-tailed gull is tipped with white). But it has been pointed out that the courtship displays of the swallow-tailed gull, easily observed on the Galápagos, differ considerably from those of Sabine's gull. Did a Sabine's flock once colonize the Galápagos and fail to migrate north? If not the Sabine's, which species was the ancestor of the swallow-tailed gull? The mystery remains.

Terns and Noddies

There are but two resident tern species on the Galápagos, the brown noddy *(Anous stolidus)* and the sooty tern *(Sterna fuscata)*. The noddy is by far the more frequently seen because the sooty tern nests only on remote Darwin and Wolf, two small islands way off the normal tourist beat. But brown noddies occur on coastal areas of pretty much all of the islands and are thus easy to see.

Brown noddies are well named. They are chocolate brown, except for a pale grayish crown. They can be observed in shallow waters feeding on fish and squid by hovering on long and graceful wings over the water and swooping quickly down to capture the prey item. They will sometimes harass brown pelicans and take fish from them.

Noddies breed on the Galápagos throughout the year and nest near the water's edge on rock faces. They do a great deal of nodding to one another during courtship, perhaps the source of the common name. In places like the bay off Rábida, snorkelers who hug the rocky face can observe brown noddies, tucked among the cliffside rocks, looking surprisingly cryptic.

Selected References

Boersma, P. D. 1976. An ecological and behavioral study of the Galápagos penguin. *Living Bird* 15:43–93.

Burtt, E. H., Jr. 1993. Cliff-facing behaviour of the Swallow-tailed Gull *Creagrus furcatus*. *Ibis* 135:459–62.

Castro, I. and A. Phillips. 1996. *A Guide to the Birds of the Galápagos Islands*. Princeton: Princeton University Press.

del Hoyo, J., A. Elliot, and J. Sargatal, eds. 1992. *Ostrich to Ducks*. Vol. 1, *Handbook of Birds of the World*. Barcelona: Lynx Edicions.

————, eds. 1996. *Hoatzin to Auks*. Vol. 3, *Handbook of Birds of the World*. Barcelona: Lynx Edicions.

Harris, M. 1974. *A Field Guide to the Birds of Galápagos*. London: Collins.

Harrison, P. 1983. *Seabirds: An Identification Guide*. Boston: Houghton Mifflin Company.

Nelson, B. 1968. *Galápagos: Islands of Birds*. New York: W. Morrow and Company.

Tindle, R. 1984. The evolution of breeding strategies in the flightless cormorant *(Nannopterum harrisi)* of the Galápagos. *Biological Journal of the Linnean Society* 21:157–64.

9

One Small, Intimately
Related Group of Birds

On the Galápagos Islands the total bird list, land and water birds, is 140 species. The 140 species include vagrants, some of which have been recorded but a single time on the islands. Of the land birds (but including rails, crakes, and moorhens), 51 species have been recorded on the Galápagos Islands and 13 (25 percent) are listed as vagrants. Eight species are migrants, and the rest are residents. The residents include 1 endemic rail, 1 endemic hawk, 1 endemic dove, 1 endemic martin, 1 endemic flycatcher, 4 endemic mockingbirds, and 13 endemic Darwin's finch species, a total of 22 endemic species, or about 43 percent of the total recorded land bird species. The remaining resident species, the paint-billed crake *(Neocrex erythrops)*, the common moorhen *(Gallinula chloropus)*, the smooth-billed ani *(Crotophaga ani)*, the dark-billed cuckoo *(Coccyzus melacoryphus)*, the barn owl *(Tyto alba punctatissima)*, the short-eared owl *(Asio flammeus galapagoenisis)*, the vermilion flycatcher *(Pyrocephalus rubinus)*, and the yellow warbler *(Dendroica petechia aureolla)*, each have mainland populations, and three (the two owls and the yellow warbler) are recognized as distinct Galápagos subspecies. The ani is the only one among these birds that was introduced. Presumably, the others arrived by their own power. So, excluding vagrants, 58 percent of the land birds regularly found on the Galápagos (including migrants) are endemic to the islands.

Now to put this Galápogeian land bird diversity in some sort of perspective, let's look at what mainland Central and South America have to offer as potential avian colonists. These areas are geographically nearest to the Galápagos, and it was recognized long ago by ornithologist John Gould and Charles Darwin that the Galápagos avifauna bore an unmistakable similarity, an affinity, to birds found on these continental areas. The possibilities for colonizing species, at first glance, seem vast. Of the approximately 9,600 species of extant birds in the world, 3,751 are found in the neotropics, the tropical region of the Americas. Of this total, if you presume that all of the 4 mockingbird species on the Galápagos are descended from a single ancestral immigrant population, not an unreasonable assumption, and if you also presume that the 13 species of the Darwin's finches are also descended collectively from a single ancestral population, an equally robust assumption, then something like 7 species of a possible 3,751 (a rail, a hawk, a dove, a martin, a flycatcher, a mockingbird, and a finch) gave rise to the present range of endemic land bird species on the Galápagos. Put another way, the endemic Galápagos land bird species evolved from something like 0.6 percent of the possible range of colonists to the islands.

Of course we are playing very loosely with numbers here. Many Neotropical birds are basically sedentary and nonmigratory and live far from coastal areas, thus not at all likely to be caught up in any kind of wind or other means of relocating from, say, the Amazonian lowland rainforest or dry interior cerrado to some remote volcanic archipelago 600 miles (965 km) from the equatorial coast. And who knows how many vagrants arrived and perished? The islands would certainly prove to be anything but hospitable to many, many Neotropical species.

The noted evolutionist George Gaylord Simpson once described what he called a "sweepstakes route" for evolutionary success, a kind of lottery where each ticket has little chance of winning, though many tickets might be purchased. This metaphor was meant to help describe evolution subsequent to island colonization. Simpson's sweepstakes metaphor translates to many species blown to sea, few actual arrivals to a distant island, and fewer still emerging as real evolutionary winners. In this game, as applied to the Galápagos Islands, Darwin's finches are the big winners, having arrived, thrived, and diversified, and the mockingbirds are running a distant second. The rail, hawk, dove, martin, and flycatcher all deserve honorable mention.

Evolution is ongoing. Once a population is geographically isolated from its parent population, it will tend, over time, to diverge genetically. It may diverge sufficiently to become recognized as a distinct subspecies (or race), and with additional time, it may eventually attain full species status by evolving to the point where it could not or would not mate and produce fertile offspring with a member of the parent species.

Both owl species have distinct Galápagos subspecies. Given the geographic isolation that these populations have from any mainland populations, chances are pretty good that they will eventually evolve to the point where they will deserve recognition as separate species, and two more Galápagos endemics will have been added. Then there is the Galápagos martin, until very recently considered an endemic subspecies of a widespread Neotropical species, the southern martin *(Progne elegans)*. And the yellow warbler, which is also considered an endemic subspecies, although the same subspecies is present on distant Cocos Island, a small island off the coast of Costa Rica. The martin just made it into the species category, and the yellow warbler may, in time, be recognized as an endemic full species. This is evolution in action.

What is fascinating to contemplate about the Galápagos land bird assemblage, a really good parlor game for ornithologists, is who could have predicted just which species would colonize in the first place and which would subsequently diversify? Could anyone have predicted it would be a finch that would be the big winner? And which finch? We tackle that question later.

All tyrant flycatchers are in a single family, the Tyrannidae, one of the most diverse avian families in the world. Tyrant flycatchers are confined to the New World and make up a grand total of 393 species. Two, just 2, are on the Galápagos and represent 0.5 percent of the total species. Many of the tyrannid species are migratory, and at least one, the eastern kingbird *(Tyrannus tyrannus),* has been recorded on the Galápagos as a vagrant. But only 2 tyrant flycatchers have succeeded on the islands (who knows how many have involuntarily "tried"), one of which has evolved away from its ancestral species to become a Galápagos endemic, the other of which remains the same species as that found widely on the mainland, though some regard it now as an endemic subspecies. Looking carefully at the meager Galápagos land birds, it might be a surprise to learn that the rail, the hawk, the owls, the dove, the cuckoo, the flycatcher, the martin, the yellow warbler, and most of the other nonendemic resident species all make sense as likely colonists. Oddly, it is the finches and the mockingbirds that seem less likely to have made the long journey to the islands. They are the real accidents of avian fate. And they are the very ones that became the most diversified. And that odd fact may not be all that hard to explain.

But first, what characteristics ought birds to have to be likely colonists? Why, unlike the finches and mockingbirds, have other bird colonists not diversified? Why did the ancestors of the present Galápagos birds immigrate to the islands in the first place?

Many years ago, in trying to explain why avian rarities show up regularly (not as much of a contradiction in terms as it first sounds), the noted field ornithologist Ludlow Griscom is reputed to have said something like, "Birds have wings and they sometimes use them." But some use them more than others. There are some species that are much more likely to show up on islands than other species. Ecologists like to talk about a theoretical notion they call *assembly rules* for how ecological communities come together—or assemble themselves. In ecology, as in most of field biology, you should apply the word *rule* with extreme caution. *Guideline* is a more appropriate word.

Be that as it may, in 1975 the biologist Jared Diamond helped focus ecologists on just what sorts of animals are most likely to be island colonizers and which are not. Some obvious points need to be considered. The first is the distance of the island(s) from the mainland source of colonizers: the closer the better. Second is the size of the island(s) and the diversity of habitats offered to potential colonizers: the bigger, the more diverse, the better. These points may seem to fall into the category of "no-brainers," but there is more to consider.

The starting point for Diamond's analysis of the colonization problem was seminal work performed by Robert MacArthur and E. O. Wilson, published in 1963 as *The Theory of Island Biogeography*. MacArthur and Wilson took the two qualitative statements above, distance from source and island size, and quantified them in such a way as to be able to predict the equilibrium number of species for a given taxon on a given island. The mathematical model stated that the number of species (of birds, for example) found on an island, any island, approximately doubles for every tenfold increase in area. So an island that is ten times the size of another has about twice the number of species. Factored into the equation for the model was a constant determined by the distance of the island from the mainland, the source of potential colonizing species. The model, when applied to a wide variety of islands, including the Galápagos, worked rather well. Looking beyond the model, what matters to the would-be colonizer is finding the island(s), the wealth of habitats offered, the climatic variability, the food availability, the lack of predators, and competing species.

Jared Diamond categorized island colonists as falling into various categories that he called *tramp species*, including such entities as *supertramps*, species with wide habitat tolerances, high reproductive potential, and high dispersal powers. What it all comes down to is (1) can the bird get there, (2) even if it can, is it likely to get there, and (3) can it survive if it gets there? And successful colonization depends on one other thing: dumb luck. A bird might get there but perish just because it is unlucky. Or it might be able to thrive on the island but never get there. That's why there are no rules, only guidelines. Nature does roll dice.

Based on the preceding three questions, what birds are most likely to colonize the Galápagos? Would-be colonizers ought to be species with wide geographic ranges, probably migratory species, because they are in the air a lot and often take the wrong direction. There is some evidence suggesting that some migrants are misdirected due to genetics because mutant genes affect the birds' innate sense of direction. Such mutations may prove adaptive in some cases if the vagrant colonizes a new area and thrives. But more sedentary species can also be tramps. Consider how weather patterns are influenced during El Niño years. Strong winds blowing off the western coast of South America could certainly misdirect resident coastal and riverine forest birds toward the Galápagos. And what would the birds find? Plenty. Recall that during El Niño years, when abnormally abundant rainfall occurs, the vegetation becomes lush, there is a marked insect flush, and, for colonizing land bird species, both animal and plant food are present in abundance, a nice welcome indeed.

If a bird species is wide-ranging, it probably has a large population and thus ample "potential lottery tickets" in the form of lost fliers looking for a place to land. The colonizer should not be too ecologically specialized but, instead, rather catholic in habitat and food choices, thus able to make do with whatever is found on the islands. If the island is young and relatively depauperate of species, almost any tramp species can have a go at the island and quite possibly succeed. But if the island is old and full of species and lots of predators and competitors (what MacArthur and Wilson would consider to be at "equilibrium"), immigrants have a more difficult, if not impossible, chance of becoming established.

Now on to the Galápagos and a look at the lucky winners.

The Crake

We begin with a bird that looks like it can barely fly, but does, and actually rather well. The Galápagos crake *(Laterallus spilonotus)* is easily overlooked or, if seen, confused with a mouse. This tiny brownish-black rail is only about 6 inches (about 15 cm) long and scoots along secretively through the wet meadows in the highland zone of most of the large islands. The bird is so reluctant to fly that it actually makes small "runways"—trails through the dense wet vegetation that it inhabits. It bears a very close anatomical and genetic relationship with a widely distributed species, the black rail *(L. jamaicensis),* found throughout coastal North America, the West Indies, and southwestern South America. Rails (family Rallidae) look awkward in flight, their seemingly small wings holding up their dumpy bodies, but many are, in fact, long-distance migrants. And despite its apparent reluctance to take to the air, this ultradiminutive Galápagos crake has obviously colonized widely: Santa Cruz, San Cristóbal, Pinta, Fernandina, Isabela, Santiago, and Floreana. It tolerates a range of habitat types and was once found both in lower, coastal elevations, among the mangroves, and the humid, wet uplands where it is now confined and considered potentially threatened by loss of habitat.

As with all of the species discussed thus far, it is really not known when the bird reached the islands, but it was sufficiently long ago that the Galápagos population now ranks as a separate species. Still, there is no indication that the various island populations are diverging genetically from one another, and perhaps the birds still move readily among the islands. Or perhaps there has simply not been sufficient time for interisland divergence to occur.

The Paint-Billed Crake and Common Moorhen

Two other members of the secretive Rallidae family are residents on the Galápagos, the colorful paint-billed crake *(Neocrex erythrops)* and the less colorful but still nice common moorhen *(Gallinula chloropus)*. Both the paint-billed crake and common moorhen species also reside on mainland South America and undoubtedly colonized the islands from there, perhaps recently. The paint-billed crake, like the Galápagos crake, is reclusive, is "thin as a rail," and scampers quickly out of sight among the reedy vegetation, but if seen even briefly, it can be readily identified by its bright orange legs and beak. It inhabits meadows and farmlands on Santa Cruz, San Cristóbal, Floreana, and Isabela. The common moorhen is more easily seen because it swims, ducklike, in open areas such as El Junco Lake in the uplands of San Cristóbal. It also frequents lagoons on most of the other major islands.

The Hawk

The tamest hawk in the world is endemic to the Galápagos Islands, and it bears the appropriate name of Galápagos hawk *(Buteo galapagoensis)*. It is found on most of the islands and is so unusually tame that Darwin remarked that he pushed one off a tree with the muzzle of his gun. The hawk is rather uniformly brown with pale underwing markings visible in flight, but identification is not a problem because there is no other hawk with which to confuse it. The hawk, true to form in tramp species, has adapted to a wide dietary range. It is a predator of everything from rats to young marine iguanas to booby chicks, while at the same time readily scavenging sea lion afterbirths and various animal carcasses. Noting its tendency to scavenge, Darwin called the bird a "carrion-buzzard."

The Galápagos hawk is a soaring hawk, a "buteo" in the parlance of birders, with wide wings adapted to permitting the bird to soar on thermal updrafts rising from the ground. It is easy to see that this species can move readily from one island to the next, and thus it should be of little surprise that it shows no interisland variation. Hawk genes probably mix fairly readily among islands. Thermal updrafts, while common over land, are hard to come by over vast tracts of open ocean, and for that reason, migratory buteo hawks strongly tend to remain over land in their journeys between breeding and wintering grounds. Still, it is a mere 600 miles (965 km) from South America to the Galápagos, and should several migrating hawks be blown off course, or should a storm displace resident hawks from the coastline, it would not be too much of an ordeal to make landfall on the Galápagos. But which hawk was the founding species? There are several candidates.

There is a hawk species that is both abundant and broadly distributed throughout the Andes ranging the length of western South America named the red-backed hawk *(B. polysoma)*, and there is a species found throughout western Mexico and through northern South America named the white-tailed hawk *(B. albicaudatus)*. Both of these species have been suggested as possible ancestors of the Galápagos hawk, and all three of the species are sufficiently similar anatomically and genetically to constitute what evolutionists call a "superspecies," a group of recently diverged species that remain, for the moment, genetically very similar.

Finally there is the Swainson's hawk *(B. swainsoni)*, which nests throughout the prairie regions of North America and migrates to spend the winter months in South America. Hundreds of thousands of these birds soaring in huge flocks funnel through the Isthmus of Panama every fall and every spring during their long-distance migration. It is possible that at some time in the past some of these migrant hawks were blown off course and ended up as the founding population of what evolved into the Galápagos hawk. Recent genetic work conducted by Rob Fleischer strongly points toward the Swainson's hawk as the best candidate for the honor of ancestral species. Based on both mitochondrial DNA comparisons as well as one nuclear gene, the Galápagos hawk and Swainson's hawk are extremely close, what geneticists call *sister species*. The mitochondrial DNA of these two species is so similar that they are in the range of what are normally "within species" comparisons.

Owls

Unlike the hawk, there is utterly no doubt which species gave rise to the two owls on the Galápagos because neither Galápagos population has as yet attained full species status. Both are subspecies, considered similar enough to their parent species that a bird of either subspecies, somehow transported back to the mainland, could mate with another of its species (though of a different subspecies) and produce fertile offspring, a kind of avian interracial marriage. Subspecies designation, like so much else in taxonomy, is a judgment call. We make our best guess and go with it. The two Galápagos owls are the barn owl *(Tyto alba punctatissima)* and the short-eared owl *(Asio flammeus galapagoensis)*. The third name of the scientific moniker is the subspecific name.

Barn owls, with their ultrasoft plumage and wide heart-shaped faces, are elegant birds. They are also good colonizers and have evolved globally into fourteen species within the genus *Tyto* and two species, called bay-owls, within their own genus *Phodilus*. One of these species, the common barn owl *(T. alba)*, occurs on all continents but Antarctica and has diverged into thirty-one recognized subspecies, including the one on the Galápagos Islands. *Tyto alba* is a good tramp species indeed. The Galápagos birds are somewhat smaller in body size and darker above than typical barn owls. Barn owls are on most of the large islands, where they occupy a range of open habitats from lowlands to highlands. As the name implies, these birds readily roost and nest within buildings, but they also use tree cavities. They feed on small mammals and insects—they take whatever they can catch. Essentially nocturnal, they perch at night on such objects as fence posts. During the day they often roost in lava tubes.

The short-eared owl, another tramp species of sorts, gets around the world pretty well. There are currently ten subspecies ranging over all continents but Australia and Antarctica. The Galápagos subspecies, found on virtually all of the islands (usually in the highlands), is considerably darker than most of the other subspecies, especially on the face and upper body. Short-eared owls are partly diurnal, frequently seen in daylight hours coursing low in erratic flight, resembling large moths as they pursue prey. They may hover momentarily before dropping to capture a mammal or, in the case of the Galápagos birds, a fledgling storm petrel.

An interesting temporal segregation of sorts has developed on some of the islands where Galápagos hawks also occur. Short-eared owls become strictly nocturnal, leaving the hawks to hunt alone in daylight. There is almost no question but that such behavior is anything but voluntary on the owls' part. The hawk, considerably larger than the owl, and with pretty much the same dietary preferences, probably intimidates any day-flying owl it sees.

The Dove

After his return to England, Darwin became a dedicated pigeon fancier, breeding many kinds of pigeon varieties as he attempted, with considerable frustration and rather limited success, to come to grips with genetics. Darwin's pigeon work is described in rather extreme detail in the first chapter of *On the Origin of Species*. His point, of course, was to show how easy it was to breed very different looking forms, each diverging from an original type. He used this analogy to show that nature could also "select" in much the same way as the breeder. I sometimes wonder if his fondness for this group of cooing birds might have had its roots in his Galápagos visit, when he made the acquaintance of the splendid little Galápagos dove *(Zenaida galapagoensis)*. This species, like virtually all other Galápagos birds, is tame and easily seen. It is one of the prettiest of the New World doves, mostly rich chestnut brown, with an iridescent green sheen on the face and neck and a distinctive blue eye ring. Many who see this chunky little dove for the first time think it resembles a quail as it strides along.

Doves are superb fliers, and a flight of a mere 600 miles (965 km), especially with a tailwind, would be no big accomplishment. Apparently that happened. The Galápagos dove, common throughout the archipelago, bears the indelible taxonomic imprint of its Neotropical origin. It belongs to a genus that contains six species of Neotropical doves, some of which are known for their far-flying habits. Most of them are found in the West Indies, though one, the eared dove *(Z. auriculata)*, is very widely distributed throughout South America and is certainly a possible candidate as the ancestral species of the Galápagos dove. But so are some others. Because doves are such energetic fliers, I think it is unlikely that the Galápagos dove will speciate further than the single species now present. It is very easy for these birds to fly among the various close islands, and thus I doubt that there is sufficient opportunity for the kind of geographic isolation necessary for genetic diversification and eventual splitting of the species.

The Dark-Billed Cuckoo

The dark-billed cuckoo *(Coccyzus melacoryphus)* is very widespread, occurring throughout the neotropics, with populations all along the west coast of Ecuador and Peru. It inhabits mangroves and coastal forests as well as riverine areas, ideal habitats in which to get caught up in some sort of storm or other event that takes it from the coast to the Galápagos, some 600 miles (965 km) away. Like doves, many cuckoos are very good fliers and could make the 600-mile journey to the islands with little difficulty.

The cuckoo is not easily seen. Unlike other Galápagos birds, it remains, like cuckoos in general, a skulker, keeping well within the dense vegetation. Most birders see the bird as it flies quickly from one clump of vegetation to another. It occurs on the major islands. In my opinion, the Charles Darwin Research Station, on Santa Cruz, is one of the most reliable places to see it.

The Yellow Warbler

The well-named yellow warbler *(Dendroica petechia aureolla),* even more than the cuckoo, is extremely widespread, occurring throughout the Bahamas, West Indies, most of North America, and South America, all the way to central Peru. On the Galápagos it ranks as one of the most abundant land birds, occurring on all islands and in all ecological zones. I don't think any other Galápagos land bird species can make that claim. Yellow warblers can seem almost ridiculously common. On the mainland, the species strongly favors wetlands, especially coastal wetlands, and thus is an ideal potential tramp species, for the same reasons that the dark-billed cuckoo is one. On the Galápagos, it lives pretty much anywhere, occupying all ecological zones. A subspecies, arguably a full species, found throughout tropical America is the mangrove warbler, a name descriptive of the bird's habitat. It is quite likely that this subspecies could have been the original colonizer on the Galápagos, and upon settling, it would have encountered mangroves, the ideal habitat for it. Soon it spread to other habitats, encountering no competitors (save, perhaps the warbler finch). And it lived happily ever after.

The Martin

The Galápagos martin, as mentioned earlier, is derived from the southern martin, a migratory species. It is also widespread and restless, occurring normally from coastal Peru to northern Chile but wandering to such far away localities as Panama, Florida, the Falkland Islands, and, at least on one occasion, the Galápagos Islands. The American Ornithologists Union has recognized the southern martin as *Progne elegans* and, in doing so, elevated the Galápagos martin to full species status, *P. modesta*. It's a judgment call. Darwin was right. Categorizing species is a slippery task.

Galápagos martins are found on most of the islands but never seem to be very abundant, nothing like yellow warblers or Galápagos doves, for example. One reason may be that their nest sites seem limited—the birds require rock crevices for their eggs. But they feed on the wing, so they can be spotted flying in pursuit of insect prey.

Flycatchers

The genus *Myiarchus* is diverse, containing twenty-two species of flycatchers, some of which are austral (southern) migrants, some of which are northern migrants, some of which are sedentary. The Galápagos flycatcher, also called the large-billed flycatcher *(M. magnirostris)*, presumably evolved from one of them. The ancestral species could have come to the Galápagos anywhere from the Bahamas, the West Indies, or the Antilles. Or perhaps the colonizer was from the scrub or mangrove forests of coastal South America. There are species of *Myiarchus* seemingly everywhere, in widely differing habitats, and they all know how to fly. More work in molecular taxonomy, where the genetic material of species is directly compared for similarity in DNA sequences, could shed as much light on the origins of this species as it already has on others (see below, the finches). Right now, the bird seems to bear its closest genetic affinities with species found in the Lesser Antilles, Jamaica, or the West Indies.

The vermilion flycatcher *(Pyrocephalus rubinus)* is a magnificent creature to behold, at least the male is. Its body is an electric red punctuated by an ebony back and wings. Seen in a certain light, males actually seem to glow. Females are subtly colored in shades of brown and gray with a hint of red on the breast and belly. The species is found mostly in the highlands and occurs on the major islands. The species is widely spread from the southwestern United States to the southern regions of South America and inhabits open areas with scattered trees. The Galápagos population has been thought by some to deserve status as a separate species, while other taxonomists classify it as a separate subspecies, *P. rubinus nanus.*

Mockingbirds

Thus far all of the birds I have described have evolved into endemic species or subspecies, an evolutionary change that distinguishes them from their counterparts, indeed their antecedents, on the mainland or from wherever they came. So the big evolutionary change for these birds was to diverge genetically from mainland populations, a genetic distance sufficient, in the eyes of taxonomists, to merit elevation to species or subspecies status for the Galápagos birds. But none of these birds has diversified *within* the archipelago. Only two of the tramps that have come to the islands have accomplished that, the mockingbirds and the finches.

There are four mockingbird species on the islands, one widely distributed, the other three limited to one island each. One, the Charles mockingbird, is now confined to but a fraction of its original range, having been entirely extirpated from Floreana. It now occurs, as an endangered species, only on tiny Champion Island, where very few of these birds remain.

Like mainland mockingbirds, the Galápagos species are members of the family Mimidae, a group, as the name suggests, known for its vocal abilities to mimic other birds. In the United States, the familiar northern mockingbird *(Mimus polyglottos)* is an example, as well as the gray catbird *(Dumatella carolinensis)* and various thrasher *(Toxostoma)* species. In general the whole group is noisy, and in a wide variety of ways.

There are seven mockingbird species in South America, and they are all ecological generalists. The two most likely ancestors of the Galápagos birds are probably the long-tailed mockingbird *(M. longicaudatus)* or the Chilean mockingbird *(M. thenca)*. Both are generously distributed along coastal South America, and transport by wind and general misdirection to the Galápagos is certainly within the realm of possibility for either species. Another contender as ancestral mocker is the tropical mockingbird *(M. gilvus)*, which ranges widely in Central America and northern South America as well as the Lesser Antilles.

However, it is important to note that in general the mockingbirds are rather sedentary. Compared with the kinds of birds discussed so far—rails, hawks, owls, doves, dark-billed cuckoos, yellow warblers, martins, and flycatchers—the mockingbird would seem less apt to be a tramp species, at least with regard to initial dispersal to the islands. I doubt anyone would, a priori, select a mockingbird as a likely colonizer. And the same, incidentally, applies to the finches. But somehow some mockingbirds made it to the Galápagos. And regardless of which species gave rise to them, the Galápagos mockingbirds have diverged significantly from the mainland group and are sufficiently distinct that they merit placement in a different and distinct genus, *Nesomimus*.

The four Galápagos species are as follows: the Galápagos mockingbird *(N. parvulus)*, the Charles mockingbird *(N. trifasciatus)*, the Hood mockingbird *(N. macdonaldi)*, and the Chatham mockingbird *(N. melanotis)*. In addition, two subspecies of the Galápagos mockingbird are recognized, the black-eared mockingbird *(N. parvulus personatus)*, found on some of the northern islands, and the Isabela mockingbird *(N. parvulus parvulus)*, found on the central and southern islands. What makes the mockingbirds unique among the examples of tramp species discussed thus far is that these birds have multiplied into additional species and perhaps continue to do so.

Each of the four species is generally similar in appearance to the others. This group has not undergone the dramatic anatomical changes from species to species so evident in the finches. There is no "vegetarian mockingbird," no "warbler mockingbird," no "woodpecker mockingbird." Though they differ from one another, a difference that impressed Darwin while he was on the islands, they are all still very much mockingbirds.

Why be so conservative? Perhaps there simply has not been sufficient time for them to diverge. But the answer may also lie in the fact that mockingbirds, including those on the mainland, are omnivores, ecological generalists that are opportunistic and thus identify a wide variety of environmental resources as food. They eat seeds, fruits, cactus pulp, insects, spiders, baby birds, eggs, whatever. My colleague Jerome Jackson observed a group of Galápagos mockingbirds feasting on a sea lion carcass, like so many little vultures. They also are frequently seen dining on sea lion placentas.

Darwin's finches are different. Whatever ancestral species founded the original population was a finch, a seed-crunching little herbivore, an ecological specialist. Finch anatomy, being what it is, would simply have to change in order to use a wider range of foods. Finches started as a more specialized kind of bird and thus were under selection pressures to diverge anatomically as they exploited a broader spectrum of foods. Mockingbirds are, everywhere, a generalized kind of bird, adapted to eat pretty much anything. In short, there is no selection pressure on them, or certainly less selection pressure, to adapt to new food sources by undergoing changes in their anatomy. So, when the mockingbirds spread to different islands, some accumulated sufficient genetic differences to isolate them as separate species, but these differences, for the most part, did not represent much in the way of structural anatomical change.

The four mockingbird species differ from one another in plumage, eye color, and vocalizations. The Charles mockingbird, now confined to tiny Champion Island, just off the coast of Floreana, is darker brownish-gray above than the Galápagos mockingbird, its eyes are reddish-brown, and it has no dark patch on its face over its eyes. The Chatham mockingbird is described as "intermediate" in plumage between the Galápagos and Hood mockingbirds, its greenish eyes are surrounded by white, and it has a brown blaze from the forehead to eye region. The Hood mockingbird, which to me is the most distinctive of the four, has a substantially more down-curved bill than the others, though otherwise it looks pretty much like the rest. Its eyes are yellowish, and there is a dark brown mark above them. And, finally, the Galápagos mockingbird has yellowish-green eyes with a grayish mark above them. So the differences are slight, not easy field characteristics, especially if you have to ascertain eye color. The birds also sound distinct from one another.

Of course, none of this is ever a field problem in identifying these birds because they never, repeat never, occur together on any single island. Each mockingbird species is entirely segregated from the others, an interesting ecological result of speciation and one quite different from the situation with the Darwin's finches. It certainly suggests that these species would be strong competitors if they found themselves together in any combination. Certainly they ought to be able to co-occur. Darwin noted that Charles Island (Floreana) is but a mere 50 miles (80.45 km) from Chatham (San Cristóbal) and 30 miles (48 km) from Albemarle (Isabela). Hood (Española) is only a 27-mile (about 4.5 km) flight from San Cristóbal. Finches have been observed over water flying between islands, but I know of no confirmed sightings of mockingbirds flying between islands. These birds have wings, but they don't use them. And, when you actually watch the birds, that becomes obvious.

Mockingbirds make short flights but not long flights. They move from bush to bush, taking wing, then alighting quickly. They like to move among branches by using their feet and legs. For example, the Charles mockingbird, when seen from our boat anchored off of Champion Island, hopped up from the dense understory and perched in a tall cactus. It was always hopping and flew a few feet at most. The story is much the same with the Galápagos and Chatham mockingbirds. They seem to use their powers of flight sparingly, almost grudgingly.

Never is this curious behavior more apparent than when watching the Hood mockingbird. This one seems even more reluctant to fly than the others. On my various visits to Española, the mockingbirds always act the same. Upon our disembarking from the panga at Hood, the inquisitive Hood mockingbirds emerge on the beach from the bordering dense vegetation. They run; they don't fly. If I sit on the beach, they approach boldly, showing no hint of fear or timidness, hopping on my legs or arms to see what might be available for caloric consumption. Bare toes are pecked, daypacks probed, all as they look upon me and my colleagues with indifferent yellowish eyes. Eventually, of course, I get up and walk away. They come after me, but they do not fly, at least not much. As I walk faster, they occasionally take wing, a brief flight, just above the ground. But mostly they run, little legs churning. It is hardly a stretch to say, at least in the case of the Hood mockingbird, that I am witnessing the first stages in the evolution of flightlessness. Now look at where Hood (Española) Island is on the map. If any of these birds gets blown off the island to the south or east, it has nowhere to go but into the sea. This low-lying, shrubby island is a dangerous place to be blown away from.

As Darwin pointed out, flight is not necessarily a beneficial trait for island-dwelling species. And the mockingbirds, first with behavior and eventually perhaps with anatomy, are adapting to that reality. I wonder what it might be like to return to Hood after a few hundred thousand mockingbird generations and perhaps be the person who describes for science *Nesomimus terrestris,* the flightless mockingbird. Remember, the cormorants pulled it off.

Darwin's Finches

Evolution's ornithological crown has thirteen jewels in it, and they reside on the Galápagos Islands. Known to science as the Geospizinae, these birds are much more widely recognized as Darwin's finches. David Lack, in 1938–1939, was the first ornithologist to make a detailed study of Darwin's finches in the field, and his work basically set much of the agenda for evolutionary biology and ecology for years afterward. After Lack's synthesis, Darwin's finches earned a place in virtually all introductory biology textbooks as the prime example of "adaptive radiation," the diversification of species from a single type. Although he was not the first to call these birds Darwin's finches, it was Lack who promoted the use of *Darwin's finches* as the popular name for the group.

More recently other researchers, most notably Peter and Rosemary Grant of Princeton University, along with many of their students and colleagues, have uncovered the phenomenal rate at which the finches respond to natural selection, as well as many other intriguing details about their remarkable ecology and evolution. The Grants' work has been described in numerous professional and popular articles that they have authored or coauthored, as well as one book by Peter and one coauthored by Rosemary and Peter. In addition at least one documentary film ("The New Explorers," with Bill Kurtis) has featured the Grants' work, and the 1994 Pulitzer Prize–winning book *The Beak of the Finch,* by Jonathan Weiner, is a popular account of evolutionary biology based largely on the Grants' finch studies.

Given the vast amount of information on Darwin's finches, in this chapter I present a mere summary of the high points, a primer on what everyone ought to know about the crown jewels of evolutionary biology.

The thirteen currently recognized species of Darwin's finches are divided among five genera. As a curious point of history, John Gould also classified these birds into thirteen species, but not the same ones recognized today. The number thirteen is a coincidence. Two genera contain but a single, specialized species: the vegetarian finch *(Platyspiza crassiostris)* and the warbler finch *(Certhidia olivacea).* One genus contains the woodpecker finch *(Cactospiza pallida)* and mangrove finch *(C. heliobates),* each quite similar to the other. There are three species of tree finches, each in the genus *Camarhynchus,* the large *(C. psittacula),* the medium *(C. pauper),* and the small *(C. parvulus).* The largest genus is the ground and cactus finches, the genus *Geospiza,* of which there are six species. Four of the species are ground finches: the large *(G. magnirostris),* the medium *(G. fortis),* the small *(G. fuliginosa),* and the sharp-beaked *(G. difficilis).* Two are cactus finches, the common *(G. scandens)* and the large *(G. conirostris).*

Even a novice birder can appreciate the close genetic relationship among at least some of the finch species. It was reasonable for Charles Darwin to wonder if they were not mere varieties when he first encountered them. All of the ground and cactus

Darwin's Finches

Common Name	Scientific Name	Food	Distribution	Best Seen On
Large ground finch	*Geospiza magnirostris*	Fruits, large seeds, caterpillars	Most islands	Santa Cruz, Genovesa
Medium ground finch	*Geospiza fortis*	Fruits, seeds, insects	Most islands	Floreana, Santa Cruz
Small ground finch	*Geospiza fuliginosa*	Varied: small seeds, buds, insects	Virtually all islands	Common throughout the islands
Sharp-beaked ground finch	*Geospiza difficilis*	Leaves, flowers, cactus pulp, insects	Seven islands	Genovesa
Common cactus finch	*Geospiza scandens*	*Opuntia* spp. flowers and fruits, insects	Most islands	Rabida
Large cactus finch	*Geospiza conirostris*	*Opuntia* spp. flowers, insects, fruits	Only three islands	Española, Genovesa
Vegetarian finch	*Platyspiza crassiostris*	Buds, leaves, flowers, fruits	Eight islands	Santa Cruz, San Cristóbal
Large tree finch	*Camarhynchus psittacula*	Mostly insects	Nine islands	Santa Cruz, Floreana
Medium tree finch	*Camarhynchus pauper*	Insects, nectar, buds	Floreana only	Floreana
Small tree finch	*Camarhynchus parvulus*	Insects, nectar, buds	Most islands	Santa Cruz, Floreana, San Cristóbal
Woodpecker finch	*Cactospiza pallida*	Insects: tool user	Regular on five islands	Santa Cruz, San Cristóbal
Mangrove finch	*Cactospiza heliobates*	Insects	Isabela only	Isabela at Black Beach
Warbler finch	*Certhidia olivacea*	Small insects	All islands but Daphne	Española, San Cristóbal, Santa Cruz

finches have similar plumages, the males black, the females and immatures a streaky brown. The three tree finch species all show identical patterns as well, with males having black heads and females uniformly brownish-green. The vegetarian finch male is blackish, the female brownish, much like the ground finches. Woodpecker and mangrove finches are grayish-olive, with no plumage differences between the sexes. The warbler finch is also olive-gray regardless of sex.

Cactus finch

The various finch species differ in abundance and distribution from island to island. You have to travel widely throughout the islands to see them all, and in some cases, you have to know just where to look. In general the small and medium ground finches, the small tree finch, the common cactus finch, and the warbler finch are sufficiently common and widespread that most visitors to the islands will see these species. By far it is the small ground finch that is the most abundant of the lot, pretty much everywhere. The large ground finch, large tree finch, woodpecker finch, and vegetarian finch are considered uncommon, though they are all widely distributed and are regularly found with some effort. Anyone who does some birding on the islands will see at least some of these. The large cactus finch is easy to see on Española and Genovesa but occurs nowhere else among the major islands. The sharp-beaked ground finch is common on Genovesa, can be found sparingly on Fernandina, Santiago, and Pinta, and is common on remote Darwin and Wolf, where, on the latter, it eats booby blood. (More on that later.) The medium tree finch is found only in the highlands of Floreana, and the mangrove finch, the rarest of the thirteen, is confined to some small mangrove forests on the west coast of Isabela.

Any book dealing with Darwin's finches, particularly bird guides, will compare them so that you may note the distinctions in body size and bill characteristics and thus go about the business of identifying the birds in the field. But beware. Even experts are apt to be uncertain about the identities of some of the finches they encounter. This is a very evolutionarily turbulent group, and it is important to realize that, in the field, you just have to identify some of the species as "uncertain." For example, there is much variability in bill size among several of the species, particularly the medium ground finch. Sometimes simply noting which island a bird is on helps with identification. The large cactus finches that live on Española look very much like large ground finches and can easily be confused with them, but large ground finches are not found on Española. However, the large cactus finches that occur on Genovesa look much different from the large ground finches that do occur there. So the large ground finches you find on Genovesa look like the large cactus finches you see on Española, but they are not large cactus finches because the real large cactus finches, at least the ones on Genovesa, look different. But the large cactus finches from Genovesa look very much like the common cactus finches found on Marchena, so much so that some taxonomists believe they ought to be considered conspecific. And, predictably enough, large cactus finches are not found on Marchena, unless, of course, they really are, but in the form of common cactus finches that look like large cactus finches.

Confused? Wait, it gets worse. Consider the scientific name of the sharp-beaked ground finch: *G. difficilis*. Now why is it called *difficilis?* It turns out that this is one of the most variable among the finches, and distinguishing it from the small ground finch, where both occur, is no easy matter. Some taxonomists believe that the sharp-beaked population on Genovesa is, in fact, a variant of the small ground finch. Another tricky field problem is picking out a large ground finch on islands such as Santa Cruz, where bill size of the medium ground finch ranges widely and can approach that of the large ground finch. At the tortoise pens of the Charles Darwin Research Station, a great place to see Darwin's finches, it is routine to watch frustrated birders try to be certain that they are really seeing a large ground finch and not a large-billed version of the medium ground finch. With luck, a large ground finch will eventually appear among the mixed flock of small and medium ground finches, affording a good comparison with the big-billed mediums, and the cry will go out, "Ah, now there's a *magnirostris!*"

And, to make identifications just that much more tricky, hybrids occur between such species as the medium ground finch and the common cactus finch or the large ground finch and the large cactus finch, species pairs that are placed in separate genera! Also, medium ground finches are known to hybridize with small ground finches. At the molecular level there is little correlation between plumage differences among the ground finches and genetic distinctions. The molecular data suggest frequent hybridization within Darwin's finches. To some degree at least, many of Darwin's finches still seem to be bathing in the same gene pool.

From just where did this gene pool originate?

When I first visited Lima, Peru, in 1979 I endured an overly lengthy flight via Quito and Guayauil before finally touching down in Lima literally in the middle of the night. Exhausted, I took a cab from the airport and passed an immense billboard featuring the countenance of the most sinister-looking rat I had ever seen. It was an ad for Racumen rat poison, to kill "ladrones en la noche" ("thieves in the night"). I was overly tired, and that sign, virtually the first thing I saw in Peru, made a lasting impression on me. I was taken to the pension where I was staying, woke the house up as I banged repeatedly on the door, and, once inside, quickly fell into bed. Next morning, to refresh and reinvigorate myself, I did what I always do. I went birdwatching. The first bird I found was a small black finch that was, of all things, jumping up and down on a branch on a little tree while making a buzzing sound. Yes, jumping up and down and buzzing.

This little finch, widely distributed in Central and South America, was a blue-black grassquit *(Volatinia jacarina)*. What impressed me about it, aside from its remarkable and persistent jumping (a mating behavior), was that it is the spitting image of a Darwin's finch, though I had not yet made an actual acquaintance with the latter. I said to myself, "Aha, so this is where they came from. Little jumping black finches." I made a mental note to check whether Darwin's finches jump up and down. They don't.

In evolutionary biology, anatomical similarity can easily be deceiving. Many other observers of birds have suggested the abundant blue-black grassquit as the ancestral Darwin's finch, but there have been other candidates as well. It turns out there is more than just one little black finch in all of the Americas. There is also the black-faced grassquit *(Tiaris bicolor)*, a species that ranges widely in the West Indies and Central America. Other grassquits have also been proposed as candidates. One species, not a grassquit, not even a finch, has been suggested, a little bird that does not even remotely resemble a Darwin's finch, the ubiquitous bananaquit *(Coereba flaveola)*. As I said, appearances in evolution can be deceiving, and it is not out of the question that a very different looking bird could have given rise to the Geospizinae. It is, however, less likely.

Actually there is a species that looks outwardly rather like a Darwin's finch and whose genetic material, whose DNA, seems to make it a likely ancestor: the St. Lucia black finch *(Melanospiza richardsoni)*. This species, now confined to the small island of St. Lucia, which sits in the West Indies roughly between Martinique and Barbados, is thought by some ornithologists to have once been considerably more widely distributed, making it a more likely candidate for colonizing the Galápagos than it now seems to be. Recent classifications of birds based on DNA similarity place the St. Lucia black finch as the likely sister species to the Geospizinae.

Be that as it may, the issue is as yet unsettled. Recent work on the molecular biology of the group suggests that the ancestral species was one of the grassquits, *Tiaris*, such as the dull-colored grassquit *(T. obscura)*, a species that ranges along the lower Andean slopes from western Venezuela, Colombia, and Ecuador, through parts of Peru and Bolivia, and into Argentina. The bird is considered to be generally uncommon throughout most of its range. The 4.25-inch (11 cm) dull-colored grassquit is, well, dull colored. It is perhaps as nondescript a bird as is possible. In volume two of *The Birds of South America*, Robert S. Ridgely and Guy Tudor described its plumage as "dull grayish olive, brown above, pale grayish brown below, lightening to whitish on the belly, sometimes with a tinge of buff on flanks and crissum." This seedeater's plumage is similar to the plumage of several of Darwin's finches, such as the mangrove and woodpecker finches and, most particularly, the warbler finch. This basic plumage resemblance, particularly to the warbler finch, becomes very interesting when Darwin's finches are compared using techniques of molecular biology.

Perhaps the most basic question about Darwin's finches is "are they all descended from a common ancestor?" Darwin initially thought not. It was John Gould who convinced Darwin that the finches were all closely related, including the warbler finch, the least finchlike of the complex. The warblerlike bill and the behavior of the warbler finch would thus be a result of a widespread process called *evolutionary convergence,* a process whereby a species evolves characteristics of other species not closely related to it. For example, the mammal we know as a porpoise is evolutionarily convergent with the one we know as a tuna. They are both sleek and fusiform, with dorsal and pectoral fins and tail flukes (though the tuna's tail is dorsoventral and the porpoise's is lateral). But porpoises and tunas are not genetically closely related; they are convergent. Convergence illustrates how natural selection can adapt different creatures to resemble one another when they occupy similar habitats.

Gould, in studying the finches, was only looking at morphology. Since Gould's study other researchers have pointed out that such genetically based characteristics as nest type, nest-building behavior, egg pattern, courtship displays, song, and even head-scratching behavior are similar among the various Darwin's finch species, again suggesting a common ancestry. Finally, analysis of DNA from the finches clearly shows that they are what evolutionary biologists call a "monophyletic group," a group recently evolved from a single ancestral species.

But one problem with using molecular techniques on the finches was that the group is so recently evolved, so genetically plastic, that the standard molecular techniques did not adequately separate the birds. That is one reason why it has been so difficult to ascertain just what the ancestral species might have been. But recently teams of evolutionary biologists (including Peter and Rosemary Grant) have been able to resolve the Darwin's finch phylogeny (evolutionary family tree) with clarity. A description of the techniques used, in one case mitochondrial DNA, and in another case microsatellite DNA, is beyond the scope of this book, but the results are of interest.

The molecular data strongly support the belief that Darwin's finches are a monophyletic group, all descended from a colonization of the Galápagos by (most likely) the dull-colored grassquit. A surprising result of the molecular studies (but one that was nonetheless anticipated) is that the warbler finch, the finch that looks least like a finch, is, in fact, at the base of the phylogeny. David Lack had placed the warbler finch as the "basal taxon," and the molecular data support his view, a view shared by others as well. Though the warbler finch does not look like or feed like a finch, it is still believed that the ancestral finch was, indeed, a finch. However, it soon evolved into the warbler finch, converging in anatomy and behavior with American woodwarblers (Parulidae). This was accompanied by the dramatic adaptive radiation into the two major "clades," the tree finches and the ground finches.

There are additional surprises. There is more than one kind of warbler finch on the islands. The genetic analysis suggests that there are at least two species of warbler finch (and eight subspecies), though they look essentially alike. One type of warbler finch, the M-E-G group, inhabits the islands of Marchena, Española, and Genovesa. The other group, called the Santa Cruz lineage, is found on the more central and western islands. Further, it may have taken only between a half million to a million years to evolve the entire Darwin's finch complex. Analysis of mitochondrial DNA suggests that the warbler finch split from the tree finch lineage (which preceded the split to the ground finches) about 750,000 years ago. Other types of molecular analyses support this general time scale.

What happened once the founding finches found their way to the Galápagos? What happened was that they speciated, they diversified. As noted earlier, of all the birds to colonize the islands, the only other group to accomplish diversification of species is the mockingbirds, and there are only four species of them. What is compelling about Darwin's finches, what impressed Darwin, Lack, the Grants, and everyone else who has studied them, is their diverse beaks. For birds, beaks are their hands, their tools, the way they make their livings. Big, strong, deep beaks crush hard seeds that little beaks can't open. Thinner curved beaks reach well into cactus pulp and flowers. Thin forcepslike beaks snatch insects. Sharp beaks can poke little holes into boobies, causing blood to flow. And so it goes. The heads of Darwin's finches are frequently depicted next to various kinds of pliers to show how each beak is uniquely adapted to deal with certain kinds of foods. The beak diversity is a reflection of the species diversity of the group, the group that inspired Darwin to suggest that these birds might have arisen from a common ancestor. When a kind of organism, in this case a small finch, splits into numerous species, each of which is adapted differently from the others, it is referred to as *adaptive radiation*. But what drives speciation in the first place and results in adaptive radiation in the second place?

Darwin took a stab at answering that question (not using the finches as an example) in *On the Origin of Species*. He proposed what he called the "principle of divergence," in which he argued that competition for various resources could drive populations to split into separate species, to form new branches on the bush of evolution. Darwin's reasoning was sound—similar kinds of organisms would tend to compete, and any characteristic that would enable some of the competitors to exploit other resources would therefore have selective value. Thus competition would drive a tendency for divergence. But what Darwin did not do was tie this principle in with that of geographic isolation facilitating genetic divergence. Put another way, it's one thing for competition to drive organisms to use a wider resource spectrum and another thing to stop gene flow among those organisms. Unless gene flow is stopped, or at least dramatically reduced, they will not develop sufficient genetic distinctions to be reproductively isolated from one another and thus will not speciate.

David Lack, with his work on the finches, succeeded in formulating a model for speciation and divergence that, while based in part on Darwin's views, was much more robust. Lack's work helped define the "central dogma" of evolution, allopatric speciation, where the term *allopatric* refers to populations that are geographically isolated from one another during a critical part of their evolutionary trajectories. This model was further refined and brilliantly championed by Ernst Mayr, one of the world's leading evolutionary biologists. Once they re-encounter each other, when they again share the same ecosystem, they become "sympatric." It is at this point that Darwin's notion of competition driving further evolutionary divergence between the fledgling species takes effect.

Several things are of real importance in understanding why Darwin's finches evolved as they did. First, the founding species was a finch. As explained earlier, finches are seed-crunching herbivores, avian specialists to a strong degree. Bill characteristics had to change, had to evolve, if the ancestral species was to exploit the full range of resources available to it on the Galápagos Islands. Second, should a tramp species find itself in a habitat in which there are essentially no competitors or no predators, it can undergo what is termed *ecological release* and exploit numerous kinds of resources that it would be prevented from using in a more species-rich community. Does anyone seriously believe the Galápagos woodpecker finch would have evolved into a tool-using bark prober if the islands already held woodpeckers? Is it likely that the woodpecker finch would have adapted to breaking off a cactus spine and using it to poke into crevices to spear insect grubs if woodpeckers were already doing the same thing with their bills and long tongues and doing it better? Both the woodpecker and mangrove finches often act like nuthatches, climbing about on trunks and large branches and probing for arthropods. Would they have been able to evolve such a foraging behavior if there were nuthatches on the islands? And if American wood-warblers such as the yellow warbler had already colonized the islands, would there be such a thing as the warbler finch? Unlikely, in my opinion.

It is amazing to watch the diverse feeding behavior exhibited by Darwin's finches. The warbler finch is primarily insectivorous. The vegetarian finch is one of the few species of birds to actually consume leaves, though it eats blossoms and buds and occasional caterpillars as well. The ground finches, in addition to consuming seeds, have developed into parasite removers, to the point where giant tortoises and iguanas assume certain postures to expose their necks and limbs to the birds. On distant Wolf (Wenman) Island, the sharp-beaked ground finch took this behavior one step further and now pecks into the backsides of boobies to feed on their blood, a behavior that, predictably enough, earned this population the name *vampire finch*.

How could a finch evolve to wound a seabird and ingest its blood? One scenario has to do with the fact that boobies are hosts to an abundant, particularly noxious, large parasite, the hippoboscid fly. This large black crablike creature, also called the louse-fly, can be seen cruising among the feathers of the boobies, munching upon the host at will. Boobies have been observed mutually grooming one another, perhaps as a response to irritations caused by the hippoboscid flies. The sharp-beaked ground finches actively capture hippoboscids, which is, of course, good for the boobies. But the parasites, often gorged with blood, may have inadvertently given the finches a taste for the substance, which became enhanced when the aggressive little finches accidently pierced the boobies' thin skins in searching for the hippoboscids. Thus what began as a mutualistic relationship, finches eating booby parasites, became a parasitic one, finches eating booby blood. It is, of course, unclear if this is a learned or innate behavior, but in all likelihood it is still in the evolutionary stage of a learned behavior, dependant on inexperienced finches watching older birds.

Of all the Darwin's finches, it is the woodpecker finch that is most often cited for its unique feeding behavior. It often uses a tool of its own choosing to procure grubs from within trees. Its close relative, the mangrove finch, is reported to perform a similar behavior. So-called tool using in animals is generally rare, though a number of other bird species manipulate objects to attain obvious goals (not including nest building). The woodpecker finch selects a cactus spine or sharp twig and actively probes bark, cracks, and other niches where insects may occur. Work in the laboratory on woodpecker finches has shown that they need to learn how to select and use the tools to attain satisfactory results.

Darwin's finches are a wonderful example of ecological release, with regard not only to what they actually eat (grubs, blood, ticks, cactus pulp, nectar, leaves, buds, and seeds of all sorts) but also to the range of habitats they occupy. Some, like the small ground finch, are generalists, found pretty much anywhere, while others, like the mangrove finch, are restricted to certain very specific habitat types. The finches came, and the finches conquered.

By comparison, mainland finch communities do not exhibit the range of resource usage evident in the Geospizinae. Dolph Schluter, an ornithologist who did a great deal of work on Darwin's finches, compared the Geospizinae with finch communities in Kenya. He learned that in comparison with Darwin's finches the Kenyan finches show significantly less diversity in beak size, eat a narrower range of seed sizes, and do not use the range of resources theoretically available to them. Schluter concluded that two forces, competition with such creatures as ants, rodents, and other seed-eating birds plus risk of predation (when, for example, feeding on the ground in the open), limit Kenyan finches in ways that Darwin's finches are not. And this comparison explains ecological release.

The next important factor in the evolution of Darwin's finches is that the Galápagos are an archipelago. The reality that there are numerous islands, most relatively close to one another, but not overly close, is a really big deal. In Lack's view, the archipelago nature of the Galápagos was the crucial factor in the eventual divergence of the species, the adaptive radiation itself. The finches fly, and they fly in flocks, ideal for prospective colonizers. But they don't fly like swallows or doves. Their little rounded wings are adapted for flights of relatively brief duration. What Lack hypothesized is that from the initial colonization of one of the islands, the finch population grew to the point where competition among the birds drove some to emigrate to the next nearest island. And, when that island became saturated with finches (recall that one of Diamond's criteria for a good supertramp is the ability to reproduce quickly), more emigrated to other islands. With each (perhaps reluctant) dispersal, populations of finches became geographically isolated from one another— they became allopatric. This could not have occurred were they not on an archipelago. Rapid reproduction is also important to the rapid accumulation of sufficient genetic divergence to provide a basis for reproductive isolation when the next group of colonists arrives or the colonists return to the island of their ancestors.

The separation of the islands plus the finches' general reluctance to move from island to island unless forced (or occasionally blown) combined to permit the allopatric populations to begin gradually to diverge genetically from one another. Such a divergence would be enhanced if the birds were exposed to somewhat different conditions, to different selection pressures on the various islands. It turns out that this was likely the case. While to human eyes, the arid zones of the various islands look the same from island to island, careful studies in the 1960s by Robert Bowman, who performed exhaustive research on the finches, showed that these zones are not at all the same, at least not as far as bird food is concerned. Bowman carefully measured the abundance of plants and found considerable variation among the islands, even within the same ecological zone. What this means is that, quite apart from any competition, finches would still be selected for different foraging patterns on different islands. A finch on one island would be selected to consume different foods from that on a different island. Presuming that such traits as bill dimensions are genetically determined (and they are—see below), finches on differing islands would evolve differently—they would diverge. Adaptive radiation would begin.

If the finches moved occasionally among islands, eventually the various diverging populations would re-encounter one another. When this happened, natural selection would act such that any differences in foraging patterns between the two emerging species would be enhanced because this would reduce competition that might arise among them. For example, if the two incipient species overlapped in beak size and were present in the same ecological zone on the same island, one, species A, would be selected to have smaller beaks, better able to handle tiny seeds, while the other, species B, would evolve larger beaks, better able to crack larger, harder seeds. All this means is that in species A those individuals with the smallest beaks would do better and thus have more offspring because they could eat foods of no interest to any member of species B. And, in species B, those with the largest bills would do better and thus leave more offspring because they had exclusive access to hard seeds that could not be used by either species A or other, smaller-beaked members of species B. The outcome of the competition is that interspecific competition would be steadily reduced as the two species diverged anatomically in bill dimensions. This is Darwin's notion of divergence of character as caused by interspecific competition. But note that this will not occur until there is already marked genetic divergence.

Allopatry begins the speciation process, followed by natural selection for food preferences, followed by effects of interspecific competition as animals hone finer and finer ecological gradations among the mixture of species, thus reducing competitive pressures. The more species there are, the more competition should drive them apart and make them ecological specialists of some sort. What Bowman's work showed, and what Lack had not realized, is that there are enough ecological differences among the islands to select for specialization among the finches, to allow for adaptive radiation, without any need to invoke interspecific competition. Even if they did not compete, the very ecological nature of the archipelago would have molded the finches to be anatomically distinct, at least to a degree. But, that said, it is nonetheless clear from Lack's work and that of the Grants and others that interspecific competition is very much a feature of the evolution of Darwin's finches. How is this competition evident?

There are patterns in the distributions of some of these finches, intriguing patterns. For example, the bird most similar to the sharp-beaked ground finch is the small ground finch. The only islands where the small ground finch fails to occur are Genovesa, Darwin, and Wolf, where the sharp-beaked is particularly common. On the islands where both the sharp-beaked and the small ground finch occur, they are altitudinally segregated. Now what does that suggest, if not competition? Another example: The common cactus finch is widespread but absent from Genovesa and Española, where the large cactus finch, the species most anatomically and ecologically similar to the common cactus finch, is common. On little Plaza Sur, there are three finch species, the small ground finch, the common cactus finch, and the medium ground finch. There is very little overlap in bill dimensions among these three species. Each is easy to identify. This pattern is repeated, with various finch combinations, on other islands.

The general "guideline" is that when several species of finch occupy the same ecological zone, there is little, if any, overlap in beak dimensions. This, in and of itself, would not be so remarkable if it were not for the fact that some of those same species, when by themselves on a different island in the same ecological zone, demonstrate a great range of variation in bill dimensions. Lack uncovered this pattern, noting, for example, how the medium ground finch that occurred on Daphne Major had beak dimensions that overlapped those of the small ground finch, absent from that island, and it impressed him as proof positive of interspecific competition facilitating divergence in bill dimensions. On islands with small and medium ground finches, the bill dimensions of medium ground finches did not overlap those of the smaller species.

The term *character displacement* has been used to describe how, when two similar species compete, there is an eventual enhancement of any differences between them that might lessen the competition. Biologist Peter Grant found that wherever there are sympatric pairs of finch species, they differ by at least 15 percent or more in some crucial parameter of bill dimensions, either in length, depth, or width.

So, to summarize, allopatry, brought about by the Galápagos being an archipelago, allows genetic divergence to occur among separated populations to the point where they become reproductively isolated and achieve full species status. From island to island the finches evolved initially in response to the varying environments themselves and, upon eventual establishment of sympatry, by interspecific competition with other finch species. And, at least for this group of birds, much of their evolutionary history may forever be clouded by the likelihood that the birds have hybridized extensively throughout their evolution and continue to do so today. The evolutionary bush that depicts Darwin's finches contains not only diverging but converging branches. The result is the ongoing adaptive radiation so evident today.

Though Darwin's finches are magnificent examples of adaptive radiation, they are far from the only examples. There is another archipelago, also in the Pacific Ocean but well to the north, where this process has produced even more dramatic results. The Hawaiian Archipelago has an endemic group of birds called the Hawaiian honeycreepers but which really should be called the Hawaiian finches. There are twenty-nine known species from this group, all in the subfamily Drepanidinae, some of which are extinct and some of which are regrettably endangered. Others may have existed for which there are no records. The Hawaiian honeycreepers bear names such as *kauai amakihi, nukupuu, akepa, iiwi,* and *ula-ai-hawane.* Genetic analysis has suggested that all are derived from some species of carduline finch, of which there are many possibilities. But many of these Hawaiian "finches" look nothing like finches. Adaptive radiation has carried this group well beyond what is seen with Darwin's finches. Some have long decurved bills adapting them for nectar feeding, and many have evolved very colorful plumage. Some have thin, forcepslike bills to capture arthropods, while others have variations of more traditional finchlike bills. Again, an archipelago. Again, a founding tramp species. Again, adaptive radiation.

And, to close this particular circle, there is the exception that proves the rule. It turns out that although there are thirteen species of Darwin's finches on the Galápagos, there are actually fourteen species in the subfamily Geospizinae. There is, in other words, a fourteenth species of the Darwin's finch. This bird *(Pinaroloxias inornata)* is otherwise known as the Cocos Island finch. It lives on Cocos Island, some 400 miles (644 km) northeast of the Galápagos, about 300 miles off the coast of Costa Rica. Recent molecular studies have shown unambiguously that the Cocos Island finch is a recently evolved species and likely colonized Cocos Island from the Galápagos. Variation in bill dimensions and variation in foraging behavior are both true of the Cocos Island finch. Individual birds specialize in one or a few of many foraging techniques. But the Cocos finch remains a single species. Cocos Island is but a single island, not an archipelago. With no opportunity for geographic isolation, no chance of establishing allopatric populations, there has been no speciation of the Cocos Island finch other than once, when it evolved in the first place, following colonization of the island. There will be no adaptive radiation of finches on Cocos Island.

The scenario described above for adaptive radiation includes some key elements: genetic divergence, natural selection, and speciation. These elements, taken together, comprise evolution itself. But each of them is based on two very important assumptions: (1) that natural selection is really true and (2) that key traits are really inheritable. If natural selection is a myth, or if traits such as bill dimensions are not inheritable, the whole house of evolutionary cards comes down.

The work by the Grants and their colleagues have clearly demonstrated that natural selection is a powerful influence on finch populations in the Galápagos, far more powerful, in fact, than Darwin dreamed. And the traits that shape the dimensions of the beaks of the birds are collectively a result of genetic action.

The Grants and their colleagues have made a long-term, and still ongoing, study of the finch populations on Daphne Major, a tiny island located just north of Santa Cruz and west of Baltra. This island was a wise choice for several reasons. First, not many species of finches occur here, and the island is sufficiently small so that the researchers could capture and band and measure pretty much every finch every year (and thus follow an individual bird's successes or failures in life). Second, the island is not a visitor site and is, in fact, hard to land on. This ensured that the researchers and their birds would not be bothered. During the several decades of research by the Grants, the finches on the islands have been subjected to both a severe drought and the dramatic rains that accompany a strong El Niño. What this amounts to is that the finches have been exposed to a wide spectrum of potential selection pressures. Daphne Major turned out to be an ideal site from which to document the ups and downs of various finch populations and the profound effects that natural selection exerted upon them.

It is not often that a Darwin's finch is selected for the cover of *Science* magazine but that happened for the October 2, 1981, issue. This issue featured an article by Peter Boag and Peter Grant that documented the single most dramatic natural selection event ever documented in nature. The abstract begins, "Survival of Darwin's finches through a drought on Daphne Major Island was nonrandom." Nonrandom. Interesting how so many people dismiss evolution as being unlikely because they wonder how organisms can evolve "by chance." Truth be told, natural selection, the evolutionary force that shapes organisms, that produces adaptation, that makes birds fly, cheetahs run, and whales dive, is not the product of chance. It is, indeed, the very opposite of random: It is nonrandom. Only some organisms survive to reproduce, and those survivors are *nonrandom* samples of the population.

In the *Science* article, the species in question was the medium ground finch. In the severe drought of 1977 most of the plants on Daphne Major stopped producing seeds, but the dramatic reduction in finch food was itself nonrandom. Hard seeds, indeed the hardest seeds, persisted, while smaller, softer seeds were not to be found. The result of the drought was a staggering loss of finches: from June of 1976 through January 1978 the number of medium ground finches dropped from 642 to 85, a loss of 87 percent. That is the first point to note about natural selection. It works by death. In the struggle for existence, not all survive. In this case, nearly nine out of ten perished.

But, most importantly, those that did survive were very much a nonrandom sample of the initial population. Only the individuals with the largest, deepest bills made it through the drought because only they could crack the hard seeds that represented virtually the only finch food to be had. Males, which tend to be a bit larger than females, survived best. Boag and Grant did not regard this severe selection event to be in any way "normal," instead describing it as a kind of "bottleneck" event that likely occurs infrequently in the histories of most populations. But the results were clear. In basically one generation of natural selection, the population of medium ground finches on Daphne Major was significantly and nonrandomly altered. A seemingly miniscule difference in bill size, a mere 0.02 inches (0.5 mm), was responsible for the outcome. But it made Daphne Major into an island of large-billed medium ground finches.

But nature can be fickle. In 1982–1983 one of the major El Niños in Galápagos history brought heavy rains that changed Daphne Major from an arid to a verdant island. Like the rest of the islands, Daphne Major had sprouted green, and lots of it. For the finches and mockingbirds the El Niño brought a lushness rarely experienced, and their populations skyrocketed. The workers on Daphne Major documented reproduction in medium ground finches, and some birds reproduced at the young age of only three months! Such a rapid response to conditions of plenty is suggestive of how effectively these birds may have colonized among the islands during past El Niños. Certainly the population on Daphne Major reached record levels. But again, success among the birds was nonrandom. Not all birds did equally well. There was an abundance of small seeds but a relative scarcity of large seeds, exactly the opposite of the drought conditions. The result was that selection moved the finches toward smaller bill and body sizes. Individuals with those traits survived best. The pendulum had swung back in the opposite direction.

The drought year and wet year produced opposite selective effects. What these examples show is that we need to apply such terms as *good genes* and *bad genes* with great caution. It was "good" to have genes for large bill and body size during the drought but "bad" to have those genes during the El Niño. Natural selection does not move in any particular goal-oriented trajectory. It is best thought of as genes, assembled in the form of organisms, tracking the conditions imposed upon them by the environment.

A final point, far from trivial, needs to be made. Natural selection only works on heritable traits. If beak dimensions were entirely determined by the environment, natural selection would have no significant effect over time. Each generation would begin with the same basic distribution of traits. Work by a number of researchers on the Galápagos has demonstrated clearly that the dimensions of finch beaks are largely determined by genes and not by environment. For the medium ground finch, about 74 percent of the beak depth is caused by genetic influence, and the remaining 26 percent by environmental influence. Body size is even more under genetic influence, its heritability measured at 91 percent.

Natural selection is ongoing on the Galápagos, rearranging the genes of Darwin's finches. And, as Darwin hypothesized, the cumulative effects of selection can be major. Peter Grant and Trevor Price used the data from the 1977 drought to perform some straightforward calculations, a thought experiment, that revealed how quickly a medium ground finch population could evolve into a large ground finch population. Their results showed that it would require a mere twenty selection events of the magnitude of the 1977 event to accomplish the change. The researchers concluded that if there was but one drought every decade, and if there was not strong counterselection between droughts, it would require but two hundred years for a new species, recognizably different, to evolve. Grant went on to suggest that even if their estimate was wrong by a factor of ten, it would still require only two thousand years. Natural selection is no small force. My mind's eye can envision Charles Darwin upon hearing this news. His jaw would slacken with initial incredulity, he would frown deeply as he digested the significance of the information, and then he would smile, a broad smile of total satisfaction. And he would say, perhaps to himself, "I was right."

Selected References

Boag, P.T. and P.R. Grant. 1981. Intense natural selection in a population of Darwin's finches *(Geospizinae)* in the Galápagos. *Science* 214:82–84.

Bond, J. 1985. *Birds of the West Indies*. Boston: Houghton Mifflin Company.

Bowman, R. I. 1961. Morphological differentiation and adaptation in the Galápagos finches. *University of California Publications in Zoology* 58:1–302.

Castro, I. and A. Phillips. 1996. *A Guide to the Birds of the Galápagos Islands*. Princeton: Princeton University Press.

Diamond, J. 1975. Assembly of species communities. In *Ecology and Evolution of Communities*, edited by M. L. Cody and J. M. Diamond. Cambridge, Mass.: Belknap Press.

Fleischer, R. C. and C. E. McIntosh. 2001. Molecular systematics and biogeography of the Hawaiian avifauna. *Studies in Avian Biology* 22:51–60.

Grant, P. R. 1981. Speciation and Adaptive Radiation of Darwin's Finches. *American Scientist* 69:653–63.

———. 1986. *Ecology and Evolution of Darwin's Finches*. Princeton: Princeton University Press.

———. 1991. Natural selection and Darwin's finches. *Scientific American* 265:82–87.

Grant, R. and P. R. Grant. 1989. *Evolutionary Dynamics of a Natural Population: The Large Cactus Finch of the Galápagos*. Chicago: University of Chicago Press.

Harris, M. 1974. *A Field Guide to Birds of the Galápagos*. London: Collins.

Kricher, J. 1997. *A Neotropical Companion: An Introduction to the Animals, Plants, and Ecosystems of the New World Tropics*. 2d ed. Princeton: Princeton University Press.

Lack, D. [1947] 1983. *Darwin's Finches*. Reprint, Cambridge, UK: Cambridge University Press.

Millikan, G. C. and R. L. Bowman. 1967. Observations on Galápagos tool-using finches in captivity. *Living Bird* 6:23–41.

Petren, K., Grant, B. R., and P. R. Grant. 1999. A phylogeny of Darwin's finches based on microsatellite DNA length variation. *Proceedings of the Royal Society of London B* 266:321–29.

Ridgely, R. S. and G. Tudor. 1989. *The Oscine Passerines*. Vol. 1, *The Birds of South America*. Austin: University of Texas Press.

Sato, A., C. O'hUigin, F. Figueroa, P. R. Grant, B. R. Grant, H. Tichy, and J. Klein. 1999. Phylogeny of Darwin's finches as revealed by mtDNA sequences. *Proceedings of the National Academy of Science* 96:5101–6.

Schluter, D. 1988. The evolution of finch communities on islands and continents: Kenya vs. Galápagos. *Ecological Monographs* 58:229–49.

Weiner, J. 1994. *The Beak of the Finch*. New York: A. A. Knopf.

Wilson, E.O. 1992 [new edition 1999]. *The Diversity of Life*. New York: W. W. Norton and Company.

10

Several Huge Whales

Wet landings or dry landings, there is no question but that visitors to the Galápagos see lots of shoreline as they travel among the various islands. Sometimes the landing is on a rocky area, the stark volcanic rubble creating tide pools where tiny fish, crabs, and various other invertebrates provide food for shorebirds and herons. Often the dark rocks, smoothed by constant tidal action, are coated with a slippery layer of algae, food for myriad marine iguanas and Sally Lightfoot crabs. On other occasions the visitor will tread on soft, welcoming sandy beaches where scattered holes in the sand reveal the numerous subterranean residences of ghost crabs. Other shorelines are populated by dense mangrove forests, trees bathed daily in sea water and uniquely adapted to the challenge of survival in high salinity. The Galápagos offer a fine example of the interfacing of the planet's two major kinds of habitats, land and water.

No evolutionist doubts that life on Earth had its initial stirrings in some sort of aqueous environment. Our cells can do nothing without water except wither and die. Life's biochemistry is so water-dependent that it is difficult to imagine a life form elsewhere in the universe that is otherwise. As we eventually explore distant planets and their satellites, we will search for a "water signature," which, if found, would provide a broad indication that something may be *living* there. As life proliferated on Earth, as simple bacterialike cells eventually gave rise to larger, more complex cells, and as some of these, in turn, came together in the form of multicellular organisms, land beckoned, offering new opportunities for adaptation. The plants and animals of coastal areas comprise various examples of adaptive solutions to living on the edge, between land and water.

Crabs and Other Invertebrates

Crabs abound on the Galápagos, but one in particular leaves a vivid image on all who come to the islands. There among the ebony rocks and gatherings of austere-looking marine iguanas scampers the most outrageously colored crab you could imagine. Vivid scarlet legs, bright red formidable claws, an electric blue belly, and a yellowish-orange upper body make the Sally Lightfoot crab *(Graspus graspus)* quite unmistakable, a real signature species of the Galápagos.

Sallys abound on rocky shores throughout the archipelago. They move energetically, sometimes so quickly that they seem to literally scoot across the water, the behavior that apparently earned them their popular name. They are sufficiently quick afoot that they are allegedly very difficult to catch, though no one tries because it would be an outright violation of Galápagos conservation ethics. But most of the time the crimson crabs move deliberately, slowly crossing rocks and rubble as they carefully clip clumps of algae with their wide claws and, in dainty fashion, place the algal salad in their mouths.

Like marine iguanas, the "Lightfoot brigade" is strictly vegetarian, preferring the same rapidly growing algae that sustain the numerous big reptiles. Its near-baseline position on the food chain, herbivory, is part of the reason for its abundance throughout the islands. Though the marine iguanas and Sally Lightfoots eat the same stuff, it is unlikely that they are serious competitors with one another. The algae grow with uncommon speed, replacing the daily mowing from reptilian jaws and crab claws. And the iguanas are certainly tolerant of their eight-legged crustacean colleagues. The crabs routinely walk across the sunning lizards without attracting the slightest glance from their scaly hosts.

What is most distinctive about these crabs, their vivid coloration, suggests an interesting evolutionary question: Why? Young Sallys are not brightly colored but are quite dark and cryptic. Such crypsis is usually an adaptation to avoid predation. Many animals, humans included, like the taste of crab, and it would seem obvious that, what with various herons and shorebirds around, these animals would be routine targets. So why do they become so obvious as they mature? Both sexes are equally striking, so it is not a case of selection for one sex to be brightly colored in order to attract the opposite sex. I can only wonder how the adults taste. Do they somehow acquire a repellent taste and could the bright red patterning be a form of warning coloration? Many insects, many amphibians, many reptiles, and at least one bird have evolved bright coloration in relation to their lack of palatability to predators. The bright coloration or patterning becomes an aid in helping predators "to remember" that whatever it is, no matter how much meat it appears to have on it, it tastes really bad and will make you sick, so they shouldn't eat it. And such an impression obviously has adaptive advantage to the would-be prey.

But there are many reports that adult Sally Lightfoot crabs are, indeed, eaten by a variety of predators ranging from fish to birds. Given that the crabs are ever so agile, their speed alone may be sufficiently adaptive in helping remove them from harm's way. And, should swiftness fail, they have another defense. They spit. When threatened they eject a small stream of water at the source of the threat. So in spite of their obviousness, they are not without defenses. But their harlequin brilliance remains, at least to me, mysterious.

Unlike Sallys, Galápagos ghost crabs, members of the genus *Ocypode*, remain out of sight most of the time. They are common throughout the islands on smooth sandy beaches and on beaches of fine volcanic rubble, where they dig burrows and stay in them until the sun sets. These crabs are best seen on moonlit nights when they emerge to forage. Opportunistic, they eat most any kind of animal they can subdue, including hatchling sea turtles. As the name suggests, they are pale cream colored, really quite elegant looking. Their most distinctive anatomical feature is that their eyes are at the ends of long retractable stalks.

Other crabs can readily be found on the islands. There is one species of fiddler crab *(Uca princeps)* commonly found among the mangroves and on muddy strands. This cosmopolitan group of crabs takes its name from the fact that the males have one very large claw and one small claw. The large claw is put on prominent display during courtship, presumably providing the female with some means of judging the quality of the lad's genes.

A tiny greenish-black crab, less than an inch (about 2.54 cm) in length, is common in the intertidal zone, especially in little tide pools. This animal, the Galápagos pebblestone crab (a species of the genus *Mithrax*) is a Galápagos endemic.

A bright red crab that is a species of the genus *Cancer* is one of several crab species commonly observed by snorkelers. This animal usually remains out of sight during the day beneath the submerged rubble, emerging at night to forage. The bright red coloration is likely a surprising form of camouflage for animals such as this crab. Consider what happens to light when it penetrates water. So-called visible light comes in various wavelengths ranging from red to violet. Light, of course, is electromagnetic radiation, whose wavelengths range from the very short (gamma, X-ray) to the very long (microwave, infrared, radio). The visible light spectrum is about in the middle of that range and is so called because we humans "see" it. Red wavelengths attenuate extremely quickly in water. The red crab looks red in normal light because it has a pigment that reflects red wavelengths. The pigment absorbs the other wavelengths. If there are no red wavelengths to reflect, the crab looks black. Divers who cut themselves on rock or coral debris often are amazed that they bleed what appears to be black blood. Coming out at night, when light is very low, ensures that all red wavelengths are absorbed and the crab can move easily without much chance of being seen.

Many coral reef fish are bright red, and these species typically remain under cover during daylight hours and emerge at night to feed. Several species of Galápagos fish, such as the 3-inch (approximately 7.5 cm) cardinalfish *(Apogon atradorsatus)* and the 7-inch (approximately 18 cm) Panamic soldierfish *(Myripristis leiognathos),* are bright red, demonstrating their nocturnal adaptiveness. Soldierfish also have immense eyes, another useful adaptation for swimming around after dark.

Several species of hermit crabs may be encountered in shallow waters around the islands. They are often active by day and readily seen by snorkelers. Each species uses gastropod mollusk shells (also known as *snail shells*) as a form of protection, a handy suit of armor to retreat into when danger threatens. The costs are that they have to carry a snail shell around with them wherever they go and they must periodically find new shells as they grow. Hermit crabs sometimes fight with one another for access to desirable shells. Survival of the fittest among hermit crabs has much to do with snail abundances and the subsequent availability of mobile homes.

Finally there is the odd-looking shamed-faced box-crab *(Calappa convexa)*. This is a hard crab to describe. It looks kind of like a big red potato that's been on the shelf too long. Subtidal and active mostly at night, this chunky, generously carbuncled box of crustacean is reputed to be incredibly quick at digging itself into the substrate when at risk. Perhaps it's just embarrassed at being seen.

Like crabs, barnacles are crustaceans, but they approach life very differently. Students worry over which college to select. Then they ponder which job offer to accept. These choices pale in comparison with the decision every barnacle must make: where to set—what tiny patch of planet to call their permanent home. As larvae, barnacles, like many invertebrates, are part of the zooplankton, the diminutive pelagic creatures that disperse their species and form an immense food base that helps support many marine animals. At this early stage of life, barnacles are free-swimming, though in reality they are strongly influenced by ocean currents. As barnacles develop, each must eventually make a profound decision. The larva must attach to a substrate and metamorphose into an adult. That adult is, from that point on, sessile. It will never move again, unless, of course, it sets on a boat or a humpback whale, in which case it goes where its mobile substrate goes. Needless to say, there must be a strong selection pressure on larval barnacles to choose wisely.

As adults, barnacles do not look very much like typical joint-legged arthropod crustaceans. Instead, many look more like miniature volcanoes: Their bodies are rearranged during their metamorphosis and become confined in little conelike shells of calcium carbonate that only open when they are immersed, allowing the animals' jointed legs to sweep synchronously in the water current, capturing zooplankton, including, of course, larval barnacles. The barnacles I describe here are the acorn barnacles. Other species, gooseneck barnacles, look roughly like a little egg suspended on a stalk.

Barnacles must set where there are other barnacles. They cannot be antisocial or they will be sterile and their genes will have to leave the pool, so to speak. Since they cannot move, barnacles have very long penises, probably the longest in proportion to their bodies of any organism. When the time arrives for reproduction, they "feel around" and hopefully encounter another barnacle, at which time they do the obvious. This approach to reproduction is admittedly risky, but there are lots of barnacles in the world, so it obviously works. One way evolution has adapted the creatures for their tactile sex lives is that most barnacle species are hermaphrodites, so an individual can both give and receive sperm.

On the Galápagos it is possible to see one of the larger of the world's barnacle species with the suggestive name of *Megabalanus galapaganus*. Look for this 2-inch-long (5 cm) acorn barnacle low in the intertidal zone or subtidally. It is relatively easy to find when snorkeling around rocky areas.

Should you stop to look at *Megabalanus galapaganus*, you might contemplate how much barnacles influenced the course of Darwin's life. While on the *Beagle* voyage, Darwin encountered an unusual barnacle contained within the shell of a mollusk. By October of 1846, fully a decade after returning to England, Charles was pretty much finished with his analysis of all of the *Beagle* collection save one, the odd little barnacle. The specimen, which became known to Darwin and his friend Joseph Hooker as "Mr. Arthrobalanus," proved problematic in the extreme. It just didn't really fit in among the barnacles already described by science. All Darwin wanted to do was to correctly classify it, but doing so consumed the better part of eight years of his life because he felt the need to become expert on all of the world's barnacles so as to do justice to the bizarre little creature that resided in the last of the bottles of specimens from Darwin's five-year voyage. Darwin did exhaustive reading about invertebrates and carried on extensive correspondence with the world's experts in marine invertebrate taxonomy. He worked and worked and worked on barnacles.

While doing his barnacle work, Darwin began to become increasingly ill, a malady that has long puzzled Darwinophiles and inspired intense debate. Why was he sick? Had he contracted some sort of illness on the *Beagle* voyage that was recurrent? Chaga's disease, an infection caused by a protozoan called a trypanosome has been suggested. Or was his illness psychosomatic, a rebellion by his body at what was occurring in his mind? As he came to understand the barnacles, how species are classified, the fine gradations among them, how they relate to one another, how they change, how change accumulates, Darwin fully realized the truth of evolution. After his barnacle work, to quote his biographer Janet Browne,

> He was no longer afraid of admitting that transmutation must have taken place. Indeed, he allowed himself an easy familiarity with the natural world, taking unsophisticated pleasure in thinking of himself as part of a living chain with animals, happily anthropomorphising the finer feelings of dogs and horses, and embracing the animal inside himself. His sense of God had virtually disappeared along with his daughter Anne. Man was nothing to him now except a more developed animal.

The reference to Anne is included because she died at age ten. She was Darwin's favorite of his children, and Darwin's faith, whatever it may have been at that time, was severely, even irreparably, eroded by the loss of his daughter. His profound shift in thinking was ultimately rooted in the loss of a beloved child, some curious finches, and a very thorough study of a lowly barnacle.

Sea Lions

Galápagos beaches are often littered with sea lions. Resembling a pack of laid-back dogs wearing wetsuits, these creatures seem to spend their lives deciding whether to lounge on the strand or frolic in the sea. Of course life is never quite that idyllic, but the Galápagos sea lions do seem to have it pretty good. There are two species found on the islands, and by far the most frequently encountered is the Galápagos sea lion *(Zalophus californianus wollebacki),* a subspecies of the widespread California sea lion. The other species is often called the Galápagos fur seal *(Arctocephalus galapagoensis)*, but it, too, is actually a sea lion and should probably be called the Galápagos fur sea lion. It is a Galápagos endemic.

Sea lion or seal? What's the difference? Both sea lions and seals are sea-going mammals, creatures whose antecedent genes resided in terrestrial carnivores closely related to dogs and weasels. The sea offered opportunities, and evolution did the rest. Sea lions and true seals are *pinnipeds,* a name referring to their limbs having adapted as flippers. A sea lion looks different from a true seal and is placed in a separate family, the Otariidae. Sea lions have external ear flaps, pinnae, while true seals have mere openings for their ears, nothing more. Second, true seals are rather pod-

Galápagos fur seal

like, their spindle-shaped bodies making them quite ill equipped for much mobility on land. But sea lions have strong front limbs: They can raise themselves up and waddle along with reasonable ease.

By the way, the prudent Galápagos visitor may want to keep in mind the fact that sea lions do move kind of quickly on land when they want to. Sometimes the large bulls, especially those on territory, have short tempers. It is extremely unwise to approach too closely one who happens to be in such a mood.

While I resided in California, I came to know the California sea lion reasonably well. Upon seeing the Galápagos subspecies, it was apparent that these animals are somewhat smaller than their cousins to the north, the males as much as 100 pounds (45 kg) less. A male Galápagos sea lion may nonetheless weigh upwards of 550 pounds (249.5 kg), which is lots of animal, especially when it swims up for a look into your facemask. Females weigh considerably less. Males are not only easy to distinguish by size but also by profile. Bulls have very thick necks and prominent, high foreheads, adaptations to the stresses imposed on them by the undeniable need to procreate.

Sea lions are an example of what Darwin described as sexual selection driven by male/male competition. In such a situation, the resource of most value is not food but uninterrupted access to females. You cannot succeed as a bull sea lion unless you are sufficiently strong and ferocious to stake a claim on a territory, a patch of coastline real estate that, for whatever reason, proves attractive to female sea lions, hopefully many female sea lions. And lots of males desire the same territory. So bulls fight, bite, butt, and bark at one another, frequently chasing each other into the sea and continuing the contest underwater. Old males often show battle scars on their necks from the bites of past contests. Males hold prime territories for as long as they can, which is often not very long, because the challengers keep coming. Females congregate in the best territories, giving the appearance that the beachmaster males have harems. But apparently what these boys really hold is the beach itself, not the females, because the females come and go as they please, and a given female may leave to mate with other bulls in other territories. The dominant bulls get to mate with as many females as will come to their stretch of beach, so a very successful male will invest a good many little genetic packages into the next generation of sea lions.

Bulls without territories gather in bachelor herds to await maturity, guts, more testosterone, or whatever it takes to challenge one of the big boys. It is usually safe to approach bachelor herds or various lounging females and young, but as mentioned already, it is not wise to approach a beachmaster on territory. Not wise at all. There are many bachelors near the trail where visitors walk along the cliff on Plaza Sur, and they rarely disturb the human traffic.

Males pay a price for their sexual struggles, a price measured in lifespan. In most pinnipeds, Galápagos ones included, the bulls live considerably shorter lives than the females. Perhaps the injuries they routinely sustain in contests with other males become infected, and they eventually succumb. Perhaps their high levels of stress hormones take a toll on their bodies' physiology.

Sea lion gestation requires nine months, and pups are born on a beach away from the main aggregation. Usually only one pup is born at a time, but because pups depend on female care for a long time, a given female may have two pups of different ages. Pups are gathered into crèches, the sea lion equivalent of day care facilities, where a female guards them while the others feed. Pups are discouraged from swimming early in life due to threats of predation by sharks. Beachmasters will sometimes chase sharks when the predatory fish are near the sea lion's beach.

Sea lions feed well offshore, and those swimming around the islands, in close proximity to the shore, may be just enjoying themselves. They certainly seem to be. In the Galápagos it is easy and safe to swim among frolicking sea lions, a memorable experience for snorkelers.

The Galápagos fur seal is similar in overall shape to the California sea lion, but as the name implies, it is hairier. It has dense fur, so its pelt was attractive to humans, causing extensive exploitation of this species that continued into the twentieth century. Numbers of these animals have increased following the protection that has been afforded them over the past fifty years. The fur seal is recognized not only by its pelt but also by its very large eyes, an adaptation for its nocturnal hunts for fish and squid.

Fur seals retire to shady, protected areas for most of the day because their fur makes them tend to overheat in the hot equatorial sun. Look for them in grottos and along cliff faces, always in the shade. Studies of their foraging behavior show that they are not just nocturnal—they are very nocturnal, spurning feeding on bright moonlit nights. They would certainly not overheat in moonlight, but they might make more obvious targets for foraging sharks, of which there are many among the islands. Sharks strongly tend to hunt at night.

The two sea lion species came to the Galápagos from opposite hemispheres. The California sea lion may have come from as far south as Mexico because the range of the species extends to the southern Baja Peninsula. Wherever it originated, it seems odd that the cold water–adapted animals made the passage across warm open ocean to the Galápagos, but the suggestion has been made that the journey could have occurred during glaciation, when the waters were colder and more hospitable than is the case now. The same is true of the fur seal, whose ancestral population is from the southern coast of South America. Both sea lion species are effectively cut off from emigration from the islands because these animals, adapted for colder waters, are literally surrounded by warm waters.

Wading Birds, Shorebirds, and a Duck

Coastal areas around the islands provide habitat and food for a variety of wading birds and shorebirds, many of the latter migrants to the archipelago.

Wading birds on the Galápagos consist of herons, egrets, and flamingos. Herons and egrets are both in the large family Ardidae, a group known for its powers of dispersal. No better example exists than the cattle egret *(Bubulcus ibis)*, an African species that apparently took upon itself an emigration from that continent to South America, where it arrived early in the twentieth century, spreading rapidly northward into the West Indies and then to the United States and, in 1964, to the Galápagos. Named for its behavior of following hoofed animals and feeding on insects that they stir up, the cattle egret has adapted well to the opportunities provided it by the agricultural lands on the major islands. The bird became a confirmed breeder on the Galápagos in 1986 and is now considered a resident rather than a migrant. Cattle egrets are small white herons with yellow bills. Both bills and legs are orangey during the breeding season, when the birds also acquire buff patches on their heads and on the back of their otherwise white plumage. Though the birds are common in uplands, they also frequent coastal areas, especially mangroves.

The other Galápagos waders include the great egret *(Casmerodius albus)*, the great blue heron *(Ardea herodias)*, the striated heron *(Butorides striatus)*, the lava heron *(Butorides sundevalli)*, and the yellow-crowned night heron *(Nyctanassa violacea pauper)*. Great egrets are like large versions of cattle egrets: tall, statuesque long-legged, long-necked birds with strong, straight beaks well adapted for grabbing fish. Herons and egrets wade into shallow water and patiently stalk fish and suddenly thrust their heads underwater to capture the piscine. On the islands great blue herons are slightly larger than great egrets and are predominantly grayish, with long black plumes on their heads during breeding season. Both the great blue heron and great egret frequent lagoons and coastal waters, and both are colonial. They are found on most of the islands and throughout the Americas.

One of the most widespread herons, a tramp species to be sure, is a bird that was until recently called the green-backed heron. It had the scientific name *Butorides striatus,* the same name as the striated heron and was, in fact, the striated heron and then some. The green-backed heron, as traditionally defined, included some thirty subspecies that range among all of the world's continents except, of course, chilly Antarctica. Most recently, the American Ornithologists Union has "split" the green-backed heron, elevating three of its subspecies into full species. Those species are the green heron *(Butorides virescens)*, the striated heron, and the lava heron of the Galápagos.

Just as Darwin learned from John Gould and others, it is hard to decide when to confer full species status. Taxonomists tend to oscillate between episodes of lumping, grouping recognizable subspecies together under one species, and splitting, giving full species status to distinct populations. With enhanced techniques for genetic analysis, splitting has become more prevalent, and by virtue of the recent splits, the Galápagos Islands have gained another endemic, the lava heron.

This bird looks like a melanistic version of the striated heron, a species widespread throughout Central and South America. The lava heron is found throughout the archipelago, often in mangrove areas, sometimes in lagoons, also along rocky areas. It is easily overlooked because it tends to be very cryptic, especially when hunching itself against the dark volcanic rocks. Unlike its larger cousins, the lava heron is not colonial but is a solitary nester reputed to be extremely territorial. The striated heron is similar, though less dark overall except for its black crown. It was obviously the species that evolved into the lava heron. But the striated heron is also a resident of the islands, though less widely distributed. It could, of course, be a case of separate colonizations, one long ago that resulted in the evolution of the lava heron and one more recently that is still the striated heron.

Rounding out the Galápagos herons, the elegant yellow-crowned night-heron is so called for its generally nocturnal foraging activities, though these herons are often sighted in the daytime and are common at dusk and dawn. Night-herons nest in colonies in mangroves and forage anywhere along coastlines where there are mangroves or rocky areas. The yellow-crowned population on the islands is considered a subspecies of the more widely distributed mainland species that can be found on both coasts of the Americas. As noted earlier, herons are apt to disperse widely, even though many populations of them, including the yellow-crowned night-heron, are mostly nonmigratory. But herons undergo what ornithologists term *postbreeding dispersal,* wandering far and wide, an easy way to get lost but also an easy way to find new areas in which to found colonies. The yellow-crowned night-heron is chunky, larger than the lava heron, overall bluish-gray, with a wide black stripe through its bright red eyes and, in breeding season, a yellow plume atop its head. Immature night-herons are brown with white speckling.

Surely one of the more unusual bird species of the Galápagos, an archipelago with no shortage of unusual species, bird or otherwise, is the greater flamingo *(Phoenicopterus ruber).* Visitors to the islands are amazed at how these exotic-looking creatures of the balmy tropics co-occupy the land of marine iguanas, fur sea lions, and giant tortoises. There are but five flamingo species extant on the planet, and three of them are confined to the Andean regions of South America: the Chilean flamingo *(P. chilensis),* the Andean flamingo *(P. andinus),* and the puna flamingo *(P. jamesi).* But none of these species colonized the Galápagos. Instead, the founding species was the widespread greater flamingo found from the West Indies and northern coasts of South America to much of coastal Africa, the Middle East, India, and Sri Lanka. Greater flamingos are known for their dispersal tendencies, and at least once, some found their way to the Galápagos. They occur on Floreana and Rábida, where most visitors see them, and also on Santiago, Isabela, Santa Cruz, and Bainbridge.

Flamingos, of course, are pink, but the intensity of the pigment varies, especially among greater flamingo populations. The Galápagos birds are recognizably paler than those from the West Indies, the likely source of colonizers. Such a pigment change could very well be due to diet. It has been shown that flamingos require certain kinds of invertebrates to maintain the intensity of pink in their plumage.

Long of leg and neck, these waders fly like cranes, with their necks outstretched, unlike herons and egrets, which curl their necks into an S pattern in flight. Flamingos have an unmistakable profile, whether in flight or on land.

Breeding season for flamingos on the Galápagos begins in July and continues through March. Many global populations of greater flamingos number in the high thousands, but the total on the Galápagos is somewhere between four hundred and five hundred birds. Courtship behavior, which precedes the beginning of breeding season, is a group activity (a "collective display") and can sometimes be witnessed in places such as the lagoon at Punta Cornoran.

My group and I enjoyed watching about two dozen courting flamingos clustered closely together in the center of the lagoon. All would raise their heads high, facing upward (a behavior termed *head-flagging*), some would spread their wings *(wing-salute),* and all would vocalize, a kind of trumpetlike call that seemed to us to carry rather well because we were not particularly close to the birds. Some of the birds would suddenly turn their necks and begin to preen a partially open wing, and this, too, is a ritualized behavior that is part of courtship *(twist-preen)*. What was most impressive was the march. The group would cluster together, wing to wing, turn, and march as one, soon turning again, all in step. This behavior, not surprisingly, is called *marching*.

The whole show was immensely entertaining, but of course, they were performing for them, not for us. But why? The answer has to do with how the birds adapt to the realities of their nesting environments. Water levels in the lagoons chosen for the nest colony can be capricious. The function of the collective display is to permit the endocrine systems of the participants to come into synchrony, to prepare each bird to mate, and to have the mating occur, and thus the nesting occur, in a relatively narrow time frame, when conditions are most favorable in the lagoons.

There is a common misunderstanding about evolution that is often promulgated by such activities as group displays. It is often said that such activities occur "for the good of the species" so that the species can continue. But the real reason for flamingos to march is not for the good of their species, at least not directly, but for their own good, each and every one of them. If a flamingo is to leave its genes in the next generation, if it is to be successful in the Darwinian sense, it must optimize its opportunity for reproduction. It does this by participating in the ritualized collective display. So the group activity is one of collective selfishness in that it is to the overwhelming advantage of each individual to get to the center of the lagoon and join the march.

Flamingos are filter feeders, straining the briny water for small invertebrates. On the Galápagos their principal food animals are shrimp and certain aquatic insects. They feed upside down. That is to say, the flamingo's head is held completely upside down and swept from side to side as it feeds. The hairlike lamellae that line its unusual bill act in a manner similar to the layers of fine baleen in the mouths of the great whales, to trap and strain animals from the water and allow the water to exit using the tongue as a pump. The beak of the flamingo has evolved to accommodate the reality that the upper beak becomes, functionally, the lower beak and vice versa.

In his insightful essay, *The Flamingo's Smile,* the evolutionist Stephen Jay Gould draws an important lesson from the bizarre beak of the flamingo:

> Natural selection, as a historical process, can only work with material available—in these cases the conventional designs evolved for ordinary life. The resulting imperfections and odd solutions, cobbled together, from parts on hand, record a process that unfolds in time from unsuited antecedents, not the work of a perfect architect created *ad nihilo.*

Charles Darwin saw flamingos on the Galápagos and elsewhere on the *Beagle* voyage and commented on how the animals inhabit salty lagoons and feed on "worms." But he saw them before he conceived of natural selection. He made no comment about how strange their beaks are and how their unique beak anatomy has resulted as a consequence of an unusual foraging behavior. Such examples, and there are many, now demonstrate the efficacy of natural selection.

When the flamingos colonized the Galápagos, they may have had company. There is another species that is commonly seen with them that also has West Indian affinities, the white-cheeked pintail *(Anas bahamensis galapagensis)*. But, although common in the West Indies and along the coast of northern South America, this duck is also found all along the western coast of South America, and this subspecies is somewhat more apt to disperse widely. So it is unclear exactly where the Galápagos birds came from, but the Galápagos population, which numbers perhaps a few thousand pairs, is now recognized as a separate subspecies.

White-cheeked pintails frequent lagoons, lakes, and pools and sometimes dive for food. More commonly they "dabble" along the shallow water and tip, tails pointing skyward, to reach the submerged plants that form their diet. Named for their prominent white cheeks and throat, they are overall warm brown and tan, with red on their bills. Pintails are the only resident ducks on the islands, though vagrants of two other species have been recorded. Pintails occur on all the major islands and many of the smaller ones.

Two large and colorful shorebirds are common residents on the archipelago, the American oystercatcher *(Haematopus palliates galapagensis)* and the black-necked stilt *(Himantopus mexicanus)*. Oystercatchers are found along the rocky shores, where they use their long bright red bills to extract mollusks from their shells. They also consume crabs and shrimp and other intertidal invertebrates. Stilts live in lagoons and eat a wide variety of small invertebrate prey.

The American oystercatcher is found on both coasts of Central and South America. The Galápagos subspecies is just a bit smaller and darker than the mainland subspecies. Given how Galápagos animals become lava colored, it would be interesting to see if this trend is followed by the oystercatcher. There are several related species found on rocky shores and pebble beaches that are blackish in overall coloration except for their brilliant red bills, which all eleven of the world's oystercatcher species have. Dark coloration, both for the lava heron and the oystercatcher, may aid them in approaching prey before being detected. Oystercatchers occur throughout the archipelago, usually in pairs. Sometimes they bow to one another and walk in tandem, clattering loudly, a ringing call of courtship.

The black-necked stilt is no different in appearance from mainland populations and thus may have colonized the islands relatively recently. It is widespread in the

Galápagos, always around lakes and lagoons. An elegant bird, it is black above and white below, with a long, slender black bill and very long red legs.

Other shorebirds are routinely seen on the Galápagos during migration. These include such species as the semipalmated plover *(Charadrius semipalatus)*, whimbrel *(Numenius phaeopus)*, lesser yellowlegs *(Tringa flavipes)*, solitary sandpiper *(T. solitaria)*, spotted sandpiper *(Actitis macularia)*, wandering tattler *(Heteroscelus incanum)*, willet *(Catoptrophorus semipalmatus)*, surfbird *(Aphriza virgata)*, sanderling *(Calidris alba)*, least sandpiper *(Calidris minutilla)*, and the ruddy turnstone *(Arenaria interpres)*. Migrant shorebirds routinely fly immense distances and frequently spend time on islands refueling before undertaking additional perambulations. Any given species can show up pretty much on any island. The birds are opportunists and go where they can find food. In addition to the species noted above, there are at least nine others that have occurred as vagrants. On my last visit to the islands we saw a curious shorebird on Punta Espinosa on Fernandina, and it proved to be a white-rumped sandpiper *(Calidris fuscicollis)*, a species so infrequent on the islands that it had been recorded only three times previously and never on Fernandina. Any birder visiting the islands should make it a point to study the shorebirds and be aware of possible vagrant species.

Fish

Fish swim, making them potential colonizers of waters around archipelagos. As the Galápagos have risen from the Pacific hotspot of their birth, the presence of the islands has provided opportunities for fish to colonize, just as it has for terrestrial species. But fish, in theory, have it easier. They are already in their element and need only find the islands, taking as long to do it as they require. Be that as it may, fish, like terrestrial species, are adapted to different climatic regimes. For example, coldwater fish do not normally swim into warm water currents. Along the North Atlantic Coast, it is routine to find fish from tropical waters in eddies from the Gulf Stream that have broken loose from the main current, but these warm water fish remain as long as they can within the warm water. Habitat matters as well, and bottom-dwelling fish do not routinely become open water pelagic travelers. What this translates into is a similar Galápagos colonization pattern for the fish as for the terrestrial biota, namely a predominance of species originating from the Americas. Of the approximately three hundred species of fish that can be found in the waters around the archipelago, 58 percent are species also found in the Panamic region (warm waters off western Central and South America), and only 14 percent are from the vast Western Pacific. Of course there are Galápagos endemics among the fish as well. About 19 percent of the Galápagos fish species are considered endemic.

There is no such thing taxonomically as a fish. It is a generic term that describes a vertebrate animal of some sort that is adapted to live in an aqueous environment and is not a tetrapod, a four-legged animal such as a salamander, lizard, bird (wings count as "legs" in this sense), or mammal. Fish can have jaws, as is the case with the vast majority (Gnathostomes), or be jawless (Agnatha), the case for lampreys and slime eels (hagfish). Some fish have skeletons entirely made up of cartilage, while others, most others in fact, have skeletons of bone. The two major taxa of extant fish are the cartilaginous fish (Chondrichthyes), sharks and rays, and bony fish (Osteichthyes), virtually all others. Both of these groups are well represented on the Galápagos and are easy for snorkelers to see.

Sharks inspire various measures of fear and awe. They are superbly adapted predators and have been around, looking pretty much like they do now, for nearly 400 million years. Apparently they are rather good at what they do, which is to capture and eat other animals ranging from invertebrates to fish to sea turtles to sea lions to occasional surfers. There are a dozen shark species that inhabit Galápagos waters but three kinds that are particularly common: the scalloped hammerhead *(Sphyrna lewini)*, the reef whitetip *(Triaenodon obesus)*, and the Galápagos shark *(Carcharhinus galapagensis)*, which is an endemic. The good news for snorkelers is that the number of shark attacks on persons documented from the Galápagos is less than the fingers of one hand. And there have been lots of humanity in the Galápagos waters. Swimming around the Galápagos is a far cry from swimming in such places as southwestern Australia, where sharks apparently regard human bathing beaches as potential fast-food restaurants. Many scuba divers and snorkelers routinely encounter various Galápagos shark species, each time without incident. Reef whitetip sharks, named for the conspicuous white tip on the dorsal fin, are known to approach humans closely but seem quite content just to look. Hammerheads are often seen from boats as the large (up to 12 feet, or approximately 3.6 m) animals swim slowly on the surface of the sea. Sometimes hammerheads aggregate in numbers of several dozen— a school of hammerheads—a behavior that is unknown outside of the Galápagos. Such masses have been observed near Devil's Crown, a well-known dive spot at Floreana. Again, these shark swarms seem not to bother divers.

The hammer-shaped head of a hammerhead is likely an adaptation to aid the fish in finding animals burrowed in the sediments. In general, sharks have superb sensory organs: vision, olfactory, tactile, and, one other, electrical. Sharks are sensitive to electric fields given off by animals' muscular and nervous systems and can thus detect a burrowed prey item even if they cannot see it, smell it, or touch it. On the hammerhead, electric sensors are widely scattered on the wide head. Like a living mine detector, the hammerhead shark swings its rectangular head methodically from side to side while swimming at the bottom, suddenly striking at a ray or other prey animal that may be covered by substrate.

Skates and rays are basically flattened sharks. They are adapted to burrow in substrate or "to fly" through the water using their greatly enlarged pectoral fins. Several fascinating species of rays are common in the Galápagos.

The manta ray *(Manta hamiltoni)* is the largest of all the rays, its "wingspread" the distance between the tips of its pectoral fins, reaching 25 feet (7.6 m). This amazing creature sometimes leaps from the ocean, making for one of the more thrilling moments on a Galápagos cruise. Mantas are unusual among rays in that they are pelagic rather than benthic. They feed on zooplankton and small fish found near the surface, capturing them with the aid of flaplike extensions bordering the mouth. Several sharks have also adapted to surface filter feeding, such as the large (25-foot, or 7.6 m) basking shark *(Cetorhinus maximus)* and even larger (55-foot, or approximately 17 m) whale shark *(Rhincodon typus)*. The opportunism that characterizes evolution is fascinating. In Chondrichthyes evolution, there have been three major directions: one that produced sharks, one that produced skates and rays, and one that produced an odd group of deep sea fishes, the chimaeras. In both the shark and ray groups, species have adaptively radiated to feed in "nontraditional ways," becoming pelagic filter feeders.

Other Galápagos rays are more "traditional," being essentially bottom dwellers. The spotted eagle ray *(Aetobatus narinari)* is common throughout the archipelago and easily recognized by the light spots that cover its otherwise dark fins and long tail. It is about 2 feet (0.6 m) across. The well-named golden ray, also known as the Pacific cownose ray *(Rhinoptera steindachneri),* is much like the eagle ray in size and shape but is uniformly pale yellow, quite unmistakable. Both the eagle and golden rays often swim in schools near the surface. Black Turtle Lagoon on the northern side of Santa Cruz is the best place to see the golden rays. Dozens of them can sometimes be found there.

Finally, there are three species of stingrays in Galápagos waters. The largest of the three, the round stingray (a *Urotrygon* species), is just over a yard (0.9 m) wide. These animals rarely move up in the water column; they spend the majority of their time burrowed just beneath the sediments or swimming at the bottom in search of worms and other potential food items. They are often in very shallow waters and pose a potential threat to people, especially those who wade rather than swim. There is a sharp spine at the base of a stingray's tail that, when the tail and body thrash, can easily penetrate skin and inject a poison that induces significant levels of pain. The easiest way to avoid an unpleasant encounter with a stingray, aside from remaining aboard the boat, is to shuffle your feet slowly along as you move. Doing so stirs up sediments and also stirs up any stingrays, whose usual behavior will be to slowly glide away in search of calmer waters.

Bony fish are the most diverse group of vertebrate animals in the world: They number nearly twenty thousand species. Though many are freshwater, most are marine. In the Galápagos there is an eclectic collection, including some remarkable for their shapes and colors. Most Galápagos visitors meet fish when snorkeling. The species they meet are the common near-shore types often associated with rocky areas and coral. Coral reefs are restricted on the Galápagos because the cold waters severely limit coral growth. But there are small reefs, all limited to the subtidal depths. Coral health is of concern on the Galápagos and has become one of the conservation issues of the islands.

On an average snorkel that may have you in the water for forty-five minutes or an hour, you can easily see between fifteen and twenty-fives species of fish, to say nothing of sea lions, penguins, sea turtles, and marine iguanas. Most of the fish that snorkelers see are feeding on small animals taken from rocks and coral. These fish have diverse shapes and sizes, and their basic anatomy is often reflective of how they feed. Perhaps most interesting, to those who take a close look, is the diversity of mouths that they have. Some, like the boldly patterned species of butterflyfish, the magnificent moorish idol *(Zanclus cornutus),* and the various surgeonfish, have sharply pointed mouths and elongate little snouts, ideal for poking into tiny spaces to pluck prey. Others, like the assorted parrotfish have blunt mouths, their teeth hardened to scrape algae from coral and rocks. The parrotfish are the sea's cows, the grazers of the waters, and when they are particularly abundant and feeding in a school, divers can actually hear them grinding coral or rock. Sound travels very well in the sea.

The diverse mouths of bony fish species reflect their many ways of snatching a living from the sea. The development of this diversity is fundamentally the same process, though on a larger scale, as the adaptive radiation for Darwin's finches described in the previous chapter.

It is beyond the scope of this book to detail each of the piscine species commonly encountered by Galápagos snorkelers. But some fish deserve mention, if for no other reason than their remarkable appearance. My personal favorite and, I expect the favorite of many, is a species named above, the moorish idol. I have watched this wonderfully colorful fish in the Galápagos as well as off the island of Komodo in far away Indonesia. This 9-inch (23 cm), laterally compressed fish resembles an angelfish in shape. It has two bold black vertical stripes on its body, yellow and white between the stripes, and a black and white tail, as striking a pattern as a fish can have. Its snout is adorned with a bright orange patch, and above each eye it has a tiny hornlike projection easily visible at close range. Its long white dorsal fin streams out behind it as it swims, and the entire visual package that is this animal can only be called elegant. An Indo-Pacific species, it is relatively common around many of the Galápagos Islands. One excellent place to see it is near the cliffs of Genovesa, but it can be found in many other areas around the islands.

Parrotfish are about the largest of the bony fish species commonly encountered by snorkelers, reaching lengths of up to 3 feet (0.9 m). They are diversely patterned, very colorful, and males and females of some species are quite different in coloration. The oddest looking of the lot is the bumphead parrotfish *(Scarus perrico)*, a turquoise fish that looks like it was whacked on the forehead and has a lump to prove it. The male of the steamer hogfish *(Bodianus diplotaenia)* is similarly graced with a cephalic dome, but it is not quite as pronounced as that of the bumphead.

Most snorkelers get to meet the 16-inch (40.6 cm) trumpetfish *(Aulostomus chinensis)*, an elongate fish that looks like it was grabbed at both ends and stretched out. Trumpetfish typically orient head downward in the water column and kind of float along, their long trumpetlike snout reaching into crevices as they probe for food.

The little, 7-inch (17.7 cm) Panamic sergeant major *(Abudefduf troschelii)* is named for its five vertical black stripes that adorn the sides of the otherwise silvery-yellow fish. Perhaps the strangest pattern of them all, though that is hard to judge—there are several good contenders—is the hieroglyphic hawkfish *(Cirrhitus rivulatus)*, a basslike fish that has brown stripes that trace a rather labyrinthine pattern on its body. And within the brown stripes there are smaller stripes that themselves make a labyrinthine pattern, giving the fish its common name. The 12-inch (30.5 cm) black triggerfish is as basic black as parrotfish are multicolored. Triggerfish are named for the sharp spines of the first dorsal fin, which can be retracted and exposed as danger threatens.

Some fish are common near the surface, often seen from the boats. The concentric pufferfish *(Spheroides annulatus)* is nicknamed *garbage fish* for its habit of swimming leisurely at the surface near boats. Deep within its fish brain it is apparently aware that garbage will be tossed overboard. This 15-inch (38 cm) fish is recognized by the curved pattern of broad white stripes on its upper side, the "concentric" pattern that gives it its common name. Puffers do puff. They swell up to dramatic girth by adding gases (dissolved out of the bloodstream) to their swim bladder, a hydrostatic organ evolutionarily related to lungs that enables bony fish to equilibrate their density with that of seawater and thus not sink or rise. Puffing behavior is an adaptation to avoid predation. Some puffer species, like those that serve as gastronomic delicacies in Japan, have extraordinarily potent toxins in their livers, making the consumption of such species an odd form of risk, adventure, and, if you live through it, dining pleasure.

Small, sardinelike fish, anchovies *(Anchoa naso)* and halfbeaks *(Hyporhampus unifasciatus)* are common near the surface. Anchovies occur in large schools and are frequent prey of Galápagos penguins. Halfbeaks are very slender and are sometimes called *needlefish* for their overall shape and elongate lower jaw. Halfbeaks are often mistaken for flying fish because they will occasionally jump from the water. But there are actual flying fish *(Exocoetus monocirrhus)* among the islands, and it is a treat to watch them leap from the sea and literally glide above the waves using their huge pectoral fins as wings.

At places such as Plaza Sur, where tall cliffs allow you to look down into the sea, you will sometimes see schools of the yellow-tailed mullet *(Mugil cephalus rammelsbergi)*. About 2 feet (0.6 m) in length, these streamlined fish often densely aggregate and feed leisurely at the surface, where they consume zooplankton in the rich waters around the islands.

There are game fish around the islands as well, posing a potential conservation problem. Species such as the dolphinfish, or dorado *(Coryphaena hippurus),* yellow-finned tuna *(Thunnus albacares),* and bonito *(Sarda chiliensis)* are examples of some of the fish that are popular game.

Whales, Dolphins, and Porpoises

They look like fish, but they are, of course, air-breathing mammals. They have evolved well beyond the development of seals and sea lions—they no longer come on land. They are the whales, dolphins, and porpoises.

The fossil record provides clear documentation of the heritage of the mammalian order Cetacea. We know their ancestors came from the land and were four-legged creatures. Molecular data such as DNA and protein analysis provide information as to which mammalian groups were the ones most likely to have given rise to the cetacean. Whales, dolphins, and porpoises have a very different heritage from seals and sea lions. The cetaceans most likely evolved from one large group within the hoofed mammals, the Artiodactyls (even-toed ungulates), the lineage beginning its track through evolutionary time about 60–65 million years ago, at the dawn of the Cenozoic Era. Recent work on cetacean DNA suggests that they are closest to hippopotamuses, having diverged about 55 million years ago.

The Cetacea rapidly diversified into two suborders: the Odontoceti (toothed whales) and the Mysticeti (baleen whales). Both of these suborders have species commonly encountered in the Galápagos. The toothed whales include dolphins, porpoises, and the sperm whale *(Physeter macrocephalus),* a species tied very much to Galápagos history and one that is still found in Galápagos waters. All toothed whales have conical teeth used to capture mobile prey such as squid, fish, and sea lions. The baleen whales are the largest of the cetaceans. They have evolved a unique mode of feeding that allows them to take in enormous quantities of zooplankton, such as tiny, shrimplike krill and schools of small fish. Baleen, or whalebone, is a proteinaceous substance, rather like fingernails, shaped as horny plates suspended from the jaw and acting as complex nets to capture tiny prey. The whale then uses its immense tongue to strain out the water and swallow the food. Some baleen whales are called *rorquals* and have expandable folds in their throats that enable the animals to strain prey from huge volumes of water.

Rorquals include the largest animal in the world today, and perhaps ever: the blue whale *(Balaenoptera musculus)*. This animal reaches a total length of 100 feet (about 30.5 m) and can weigh up to 160 tons (146 metric tons), which is 320,000 pounds (144,000 kg)! Blue whales are not usually found in Galápagos waters, though they range throughout the world's oceans. But other rorquals are encountered around the Galápagos. These include the fin whale *(B. physalus),* the humpback whale *(Megaptera novaeangliae),* and three other, smaller species. Fin whales can reach lengths of over 80 feet (about 25 m), and humpbacks measure up to 60 feet (about 18 m). Humpbacks are known for their dramatic breaching, energetic leaps from the water that end in a resounding splash. Rorquals, by their unique filter-feeding behavior, have effectively shortened the oceanic food chain to three steps: phytoplankton (tiny plants suspended in the water column) to zooplankton (especially shrimplike krill) to whale. This enables the giant mammals to achieve their size; the whales obtain a large share of the sun's energy as it passes through but two other groups of organisms before being claimed by the whales.

Toothed whales are generally much smaller than baleen whales. There are about ten species of dolphins that are found around the Galápagos. One of the most frequently seen is the bottle-nosed dolphin *(Tursiops truncates),* a streamlined animal that often rides the bow waves of boats moving among the islands. But the one that attracts the most attention is the largest of the lot, the orca *(Orcinus orca).* Sometimes called *killer whales,* orcas are rather like wolf packs. The orca is the largest of the dolphins, reaching 30 feet in length (about 9.5 m) and weighing nearly 10 tons (9 metric tons). They are mammals, of course, so they are relatively intelligent, and they maintain strong social bonds and stable groups. They are behaviorally complex animals, which is likely the case for mammals (and birds) in general. Pods of orcas can number up to thirty animals of various ages, though smaller pods are much more common. Adult males are easily recognized by their extremely tall dorsal fin. Orcas roam the world's oceans, usually feeding on fish, sea lions, and, if they can catch them, juvenile sperm whales. They normally hunt well offshore. The abundance of major prey types, especially fish and sea lions, in Galápagos waters is a strong attraction.

You have to be on watch when sailing among the islands. A Galápagos sail is not the time to sit and read a novel. My wife, Martha, saw, well in the distance, what she thought to be spouts, and spouts they were—from orcas! We were sailing north along Bolivar Channel from Fernandina on our way to Santiago, off Cape Berkeley, just about on the Equator. Bolivar Channel is perhaps the best place in the islands to encounter cetaceans. It is there that the equatorial undercurrent rushes against the Galápagos Platform and rises to the surface with its abundant nutrients gathered on its slow crossing from the Indonesian area. The higher productivity of these waters is evident while snorkeling around Fernandina, for example, where the water is markedly cloudy with plankton, whose populations have burgeoned from the nutrient-rich water.

The orca pod we encountered was small, a group of five individuals that included a large male. There was an immense bird flock around the whales consisting of frigatebirds (one dangling what appeared to be a long strand of intestine of some sort, perhaps from a sea lion that had been lunch for the orcas), storm petrels, Audubon's shearwaters, dark-rumped petrels, and one brown pelican. Our boat, whose entire crew was jubilant over encountering orcas, approached the active whales closely enough that we had very good views of them as they periodically surfaced. This is the sort of experience that draws you back to the Galápagos.

The Galápagos Islands served as a restocking place for whaling ships of the nineteenth century. The animal that they sought was the sperm whale. Herman Melville, when doing research for *Moby Dick*, visited the Galápagos. Fortunately the islands no longer contribute to the reduction of one of the world's most remarkable mammals. Sperm whales reach lengths of close to 70 feet (about 22 m) and can weigh up to 50 tons (45 metric tons). They dive deeply, sometimes over 3,000 feet down (about 900 m), in search of large, indeed giant, squid. Like other whales they can be highly vocal, but their odd clicking sounds function for echolocation in the profound blackness that is the deep ocean. Like orcas, sperm whales move in groups, probably families led by one dominant male. Juvenile sperm whales are prime targets of orcas, and there have been well-documented accounts of orcas isolating a young sperm whale from its pod, only to quickly subdue it, tear it apart, and eat it.

Sperm whales are frequently sighted in and around Galápagos waters, though they seem a bit more common on the western side of the archipelago. One study, which lasted 2.5 months, found about two hundred female and immature sperm whales. These animals were thought to be arranged into thirteen groups, the usual group consisting of twenty females and immatures and one to two calves.

I saw my first sperm whales when sailing from Guayaquil to San Cristóbal. The MV *Santa Cruz* was sailing through calm seas, and I was watching flying fish launching themselves from near the ship's bow, flying laterally away from the large vessel. Remarkably, they really do fly, often 100–200 feet (30.5–61 m) in the air, their stiff "wings" making them resemble seagoing dragonflies. One of the passengers saw me watching the fish and asked me what those "strange little birds" were. But the flying fish soon lost my attention as I looked out and saw whales spouting. The various whale species have distinctive spouts, helping to identify them at a considerable distance. The spout, of course, is the nostril of the animal, located on the upper side of the head, where it is well adapted to allow the beast to exhale after a dive. This spout was short and dense and angled distinctly forward and to the left, the pattern shown by sperm whales. I wanted to shout "Thar she blows!" but refrained. We passed many sperm whales that afternoon and saw juveniles and adults, some quite close to our ship. It was easy to see the huge squared off head and the bulbous compartment containing spermaceti, the oil that was once mistaken for semen and that gives the animal its name.

Sea Turtles

One final group of animals deserves mention before we leave the coasts and seas of the Galápagos. Several species of sea turtles are found in Galápagos waters. Most are rare. It is unusual to find the olive ridley *(Lepidochelys olivacea)*, the hawksbill *(Eretmochelys imbricata)*, or the immense leatherback *(Dermochelys coriacea)*.

But one species, the Pacific green turtle *(Chelonis mydas agassisi)*, is common around the islands and is usually seen at least once or twice on any Galápagos cruise.

Sea turtles are kind of the reptilian equivalent of sea lions. They belong to a terrestrial group, reptiles, which has, from time to time, branched off groups that have become marine. And they are still dependent on dry land for reproduction. In the Mesozoic Era, the age of reptiles, there were many seagoing reptiles, including odd "sea serpents," the plesiosaurs; fishlike ichthyosaurs; huge seagoing lizards, the mosasaurs; and, even then, sea turtles. Of that distinguished cast, only the sea turtles remain on the planet today.

Sea turtles are surface dwellers, ranging widely across the oceans of the world, often eating such simple fare as jellyfish. Around the Galápagos, the green sea turtle feeds mostly on algae.

Sea turtles are often fairly large, the green weighing over 200 pounds (90 kg) and sometimes as much as 300 pounds (135 kg). There is much variability in color among so-called green sea turtles. I've seen some that look green but others that are brown and almost blackish. One of the most reliable places to see them is Black Turtle Lagoon on the northern side of Santa Cruz.

Sea turtle species have become endangered in many areas around the world, but on the Galápagos, green sea turtles are still doing well. They use the islands as nesting beaches. Being reptiles, they lay eggs and must come on land to do so. A gravid female hauls herself out of the sea and, making a distinctive wavy trail in the sand, laboriously treks up the beach to above tide line. The female turns around and, using the flippers that are her hind legs, excavates a sandy pit in which eggs are laid. Eggs are generally safe unless dug up by roving pigs. Hatchlings must scamper from the upper beach to the sea, a behavior that is innate and their first real test of survival. These tiny turtles fall prey to many animals from herons to frigatebirds to dogs.

Sea turtles do roam widely in the seas of the world, and studies of the Galápagos population that involved marking individual animals have demonstrated that these creatures can swim to the shores of mainland South and Central America and show up in such scattered localities as Costa Rica and Peru.

Selected References

Browne, J. 1995. *Charles Darwin: Voyaging*. New York: Alfred A. Knopf.

Castro, I. and A. Phillips. 1996. *A Guide to the Birds of the Galápagos Islands*. Princeton: Princeton University Press.

Constant, P. 1992. *Marine Life of the Galápagos*. Hong Kong: Twin Age Limited.

Harris, M. 1974. *A Field Guide to the Birds of Galápagos*. London: Collins.

Merlen, G. 1988. *A Field Guide to the Fishes of the Galápagos*. London: Wilmot.

Orr, R. T. 1967. The Galápagos sea lion. *Journal of Mammalogy* 48:62–69.

Pennisi, E. 1999. Genomes reveal kin connections for whales and pumas. *Science* 284:2081.

Steadman, D.W. and S. Zousmer. 1988. *Galápagos: Discovery on Darwin's Islands*. Washington: Smithsonian Institution Press.

11

Galápagos 3000?

What will the present millennium mean for the Galápagos? Will the unique ecology of the archipelago endure to the next millennium, a long time in human terms but a brief instant in geologic time? Earlier in this book I state that evolution is the ultimate existential game. No life form is guaranteed immortality. No adaptation is foolproof. Conditions change. For most organisms the environment eventually deteriorates to the point where either an organism evolves to become almost unrecognizable compared with what it once was (today's birds don't really look all that much like their presumed ancestors, the dinosaurs, for example) or the life form simply becomes extinct, joining a very long list indeed.

Conservation efforts are analogous. It is possible to preserve a species or a habitat or, in the case of the Galápagos, an archipelago, to pass laws on behalf of creatures and habitat, and to enforce those laws. But, when a place is protected from potential depredations by humans, the essence of what conservation policy is, it is always at least theoretically possible to unprotect it or to flagrantly ignore the protective laws. Once a species is extinct, it is forever relegated to the past tense. But if it is extant and protected because it is endangered or is part of a national park, it may be poached, inadvertently or purposefully harmed by people, outcompeted or preyed upon by alien species, or simply declared no longer under protection. The fate of organisms and ecosystems is always subject to the whims, social pressures, and political decisions of human beings. Therefore, in any analysis of a region and its conservation policies, it is essential to consider what is, what should be, and, realistically, what likely will be.

The Galápagos became an Ecuadorian national park in 1959, approximately four decades ago, when 97 percent of the total land area of the islands was declared to be protected. This was followed in 1978 by the recognition of the archipelago as a United Nations World Heritage Site. The Charles Darwin Research Station operates under a charter from the United Nations Educational, Scientific, and Cultural Organization. In many ways Galápagos National Park Service has been exemplary in demonstrating what steps need to be taken to preserve life forms that, by their very isolation and peculiar evolution, are extremely vulnerable to invasive species and human disturbance. Human settlements and immigration were strictly limited; alien species such as goats, pigs, and dogs were pursued with the intent to eradicate; and ecotourism was carefully regulated, allowing only highly controled access to the islands.

The Charles Darwin Research Station has been pivotal in focusing serious scientific attention on the Galápagos ecosystems and in combining research with solid conservation efforts. The research station ranks with such places as Barro Colorado Island in Panama and La Selva Biological Station in Costa Rica for its contributions to understanding regional biology. One recent comprehensive study noted that over seven hundred scientific missions to the Galápagos used the Charles Darwin Research Station as their base and that the total number of scientific publications on the Galápagos now totals in excess of six thousand. Over five hundred undergraduate students from Ecuador have received training at the research station as naturalist guides and for other volunteer duties, and over one hundred scientists have received PhD degrees and/or masters degrees as a result of research on the Galápagos. Beyond that, the research station has served to educate the thousands of annual visitors to the islands, many (if not most) of whom come with relatively little solid knowledge of the remarkable area that they are visiting. Without the Charles Darwin Research Station it is doubtful at best that the native flora and fauna would be as healthy as they are today.

Consider a comparison of the Galápagos with the Hawaiian Islands. Both archipelagos are volcanic and isolated, with no historical connections to the mainland. The Hawaiian Islands are thought to be about 70 million years old, much older than the Galápagos. Both archipelagos evolved unique species assemblages such as Darwin's finches on the Galápagos and the Hawaiian honeycreepers on the various Hawaiian islands. But, in terms of land birds alone, the Hawaiian Islands have seen the extinction of more than 75 percent of their endemic species. In contrast, not one single land bird species has been lost in the Galápagos. The Hawaiian Islands have a vastly larger human population than the Galápagos, about eighty times as much. Both island groups have experienced invasions of alien species ranging from goats and pigs to numerous plants, but much stricter protection, along with vigorous eradication efforts, have occurred on the Galápagos, and consequently, the Galápagos endemics have fared better. There is no comparison with how tourism is regulated between the two places. In the Galápagos tourism is regulated rather like it is in various African national parks, but for the reverse reason. In the African parks risk of bodily harm from animals is a major reason for strictly enforced rules of conduct. There is no such risk in the Galápagos. The strict regulations are intended to protect the ecosystems and the animals, not the tourists.

The rules of the Galápagos National Park are indeed strict but are of obvious necessity to protect the fragile ecology of the islands and the extreme vulnerability of the animals, most of which show no anxieties at the close presence of human beings. No items of any sort may be removed from any of the islands, not even driftwood, a leaf, a feather, or a tiny lava rock. Likewise, nothing may be transported among the islands. No seeds, no pieces of plant material, nothing. This necessitates visitors carefully checking sneakers, boots, and clothes before disembarking to an island and upon returning to the boat. No animals of any sort may be brought to the islands.

There is no picnicking on the islands and certainly no littering or graffiti. Every item of trash, every gum wrapper, must be carried back to the boat. Galápagos animals are never to be touched, chased, prodded, or threatened, and they are certainly not to be fed. Visitors must remain on trails with the naturalist guides. No more than twenty people can be with any one guide, and the guide is responsible for keeping the group together at all times.

The strict conservation rules imposed by the national park service are irritating to some ecotourists. A friend of mine who accompanied me on one of my trips says he will never return, in part because he felt overly confined while visiting each of the islands. He expressed the opinion that touring the Galápagos was kind of like walking through an outdoor museum. Of course it's true, the islands are a museum of sorts, a living museum. But to my friend, the experience felt like overly structured tourism, not exploration.

Sometimes you want to linger, just be alone, look at the wildlife and scenery, and be contemplative. A serious photographer may want to wait for the right profile of a marine iguana, the perfect light. But that won't happen with a Galápagos tour group. You may want to wander along the shoreline looking in tide pools or scanning the rocks for rare migrant shorebirds, but the guide will tell you to stick close while he or she lectures on volcanic rocks or booby behavior. You may already know these things or be disinterested, but you are expected to remain with the group until the guide says to proceed. As an ecologist, and someone who fancies himself competent as an observer of natural history, I like nothing better than to be alone to explore and would certainly enjoy having such freedom accorded me while on the Galápagos. But I truly shudder to think what the net effect of such unbridled liberty would be if every tourist to the islands could walk anywhere and do whatever he or she wanted.

Tour groups increase annually as global interest in ecotourism grows. There is pressure to provide more boats and train more guides. The national park service attempts to strictly regulate island visitation, and each tour boat is given a schedule to which it must adhere. Efforts are made to coordinate tour boat itineraries such that tour groups do not crowd together on a single visitor site at the same time. Still, when my group visited Rábida, for example, there were several other tour groups present on the small red lava beach who were crowded around the tiny lagoon and walking the trail through the cactus forest. We politely passed one another, but the overall experience was compromised, at least somewhat, by the plain fact that there were too many human beings at the same place at the same time. I was glad that such human traffic jams did not occur when visiting sites on the other islands.

Tourism is fundamental to the Galápagos economy. Of those individuals gainfully employed on the islands, some 40 percent are working in some form of the tourism industry. A recent comprehensive report on conservation issues and human impact on the islands notes that there are now about fifteen large- and medium-sized tour vessels plus about eighty small boats, all visiting some fifty-five designated visitor sites. These vessels operate year-round. Thus far, despite the heavy human traffic on the islands, the trails are basically intact, and there is little evidence of ecosystem degradation. In other words, the visitor sites are well maintained and look appealing and natural. The strict rules work, but pressures are being exerted to allow for more vessels. Such trends will eventually place a significant strain on the existing regulations. The point here is that the highly regulated, conservation-based tourism industry is itself in potential long-term jeopardy should pressures for increasing human traffic continue. Tourism is a bit like maintaining a healthy body weight. You can weigh too little or too much. It is possible to have a site be threatened by too little tourism, which may mean there is not sufficient justification for protecting the site against other economic pressures. But it is also possible to erode the site by too much tourism. That is what now poses a problem for the future of the Galápagos.

Island wildlife, by its very isolation and long-term evolution in the absence of significant predators and competing species, is extremely vulnerable. On the Galápagos, as on Hawaii, there have been numerous examples of depredations by such species as goats, donkeys, pigs, feral dogs, and black rats. These alien animals are adapted to be ecological generalists and to reproduce quickly, two characteristics that make them very successful against the likes of giant tortoises and land iguanas. Humans also can have, to put it mildly, a significant negative effect on native flora and fauna. Human beings, the most successful single species ever to exist on Earth, are supreme ecological generalists as well as extraordinary competitors. The combined effects of human beings and various alien animals (some introduced on purpose, some by accident) have imposed significant stresses on Galápagos species, stresses that remain today and that demand constant attention.

A case in point is the history of goats on Pinta Island. In 1959 a single fisherman released three goats, two females and a male, on Pinta, an island that up to that point was goat free. The population of feral goats on Pinta was estimated to be somewhere between fifty thousand and one hundred thousand by 1970. If we assume the lower of the two estimates, goats increased by a factor of 16,667 in eleven years! Four plant species were eliminated from the island by goat consumption, and five others were significantly reduced. Goat eradication was conducted by the national park service, and the last of the goats was evicted in 1986.

Goat eradication efforts have occurred repeatedly among the islands, yet goats still thrive on five islands. On the large island of Santiago eradication programs have reduced the number of feral pigs, but the goat problem remains unchecked. So severe have the effects of pigs and goats been on Santiago that it is uncertain if the island vegetation will recover if and when the goats are finally eradicated. On most of the other islands, when goats have been eliminated, vegetation has tended to recover.

When I last visited the islands, there was a vigorous campaign in effect to reduce the thousands of goats that are on Volcán Alcedo on Isabela. Alcedo is one of the strongholds of the domed shell tortoises, accounting for 36 percent of all the tortoises present on the islands. The goats are consuming vast amounts of vegetation and thus are direct competitors with the tortoises for food. Beyond that, the changes brought about in the microclimate of the tortoise nesting grounds as a result of vegetation loss by goats could alter the sex ratio of the tortoises because the sex of a tortoise depends upon the temperature of the soil in which the egg develops.

Pigs, like goats, can convert a pristine island into an ecological disaster zone. Pigs consume large quantities of vegetation, but unlike goats, pigs are also carnivorous. They root up tortoise eggs and consume eggs and young of both marine and land iguanas and green sea turtles.

These same Galápagos endemics are prey for feral dogs. The dogs consume both nestlings and eggs but will also attack adult marine and land iguanas, boobies, penguins, and even young fur sea lions. If feral dogs, pigs, and goats were left unchecked, if nature was left "to take its course," most of the Galápagos animals would quickly become extinct. The islands must be vigorously managed if there is to be any hope of long-term survival for the endemics.

Not all alien species are as obvious as pigs, goats, and dogs. One of the worst aliens introduced to the Galápagos is the little fire ant *(Wasmannia auropunctata)*. This animal is widespread in Central and South America and is known for its severe sting. It has spread widely in the southern United States and is currently on the move in other parts of the world as distant as New Caledonia and the Solomon Islands. The first fire ants on the Galápagos showed up on Santa Cruz almost a century ago. Unfortunately, they have spread among the islands, occurring now on Santa Cruz, Floreana, San Cristóbal, Volcán Sierra Negra on Isabela, and several other, more isolated spots among the islands. They have been documented to have had a negative impact on seventeen of the twenty-eight ant taxa on the islands and to have reduced populations of two spiders and a scorpion. They have been observed attacking hatchling tortoises and the eyes and cloacal areas of adult tortoises.

Alien plants are as much a threat to Galápagos endemics as alien animals. Four plant species in particular (among the nearly two hundred species introduced by humans) have presented problems by competing vigorously with native species. These are the quinine tree *(Cinchona succirubra),* the guava *(Psidium guajava),* a shrub *(Lantana camara),* and a grass *(Pennisetum purpureum).* Each of these species can grow aggressively enough to exclude native species.

The Charles Darwin Research Station has stewarded efforts toward reducing alien species and promoting the reintroduction of Galápagos animals to various islands. The research station is best known for its efforts on behalf of the giant tortoises, but it also assists the propagation of land iguanas and works to protect the dark-rumped petrel colonies. The tortoise conservation program, established in 1965, has been successful in its efforts to rear young tortoises in captivity and to eventually repatriate them back to the islands on which they belong. Efforts on Española and Pinzón are the most noteworthy to date.

Pinzón tortoises, which are one of the saddleback races *(Geochelone nigra ephippium),* were significantly reduced by black rats that had inadvertently been introduced shortly before the twentieth century. The rats devoured hatchling tortoises, and evidence suggests that the tortoise population on Pinzón did not reproduce with any success for about seven decades. The eggs of Pinzón tortoises were carefully transported to the Charles Darwin Research Station and incubated in solar-heated wooden chambers. The young that hatched were reared in specially constructed pens, and many of the hatchlings survived. The animals, when five years old, were reintroduced back to Pinzón beginning in December 1970. These animals have generally thrived and are now attempting to breed. Rats still occur on Pinzón, and their elimination is required for there to be reasonable hope that the tortoises can endure.

Land iguanas, as noted by Darwin when he visited James Island (Santiago) 165 years ago, were once vastly abundant. The combined effects of goats, which eat vegetation that is food for iguanas, dogs, which kill young and adult animals, and

pigs, which root up the iguana nests and eggs, have been sufficient to completely annihilate land iguanas on Santiago. Land iguanas have suffered as well on other islands. The Charles Darwin Research Station rescued land iguanas that had survived
dog attacks on colonies in Santa Cruz and Isabela, an effort that helped initiate a captive breeding program similar to that for the tortoises. Some of the animals from the
captive breeding program were successfully released back on Isabela, at the colony
site on Conway Bay, beginning in 1980. The research station is essentially engaged
in an ongoing struggle to eradicate dangerous alien animals from areas where they
can threaten Galápagos endemic animals and to maintain healthy breeding animals
at the research station to supply animals for repatriation. This struggle shows no
signs of ceasing, only increasing.

The dark-rumped petrel breeds in highland colonies on Floreana, San Cristóbal,
Santa Cruz, and Santiago. These colonies are but a fraction of the numbers of
this species that the islands once supported. The dark-rumped petrel has been vic-
timized not only by goats, pigs, and black rats but also by humans. The breeding
birds locate their nests in burrows at upper elevations on the major islands, and
many colonies were destroyed by conversion of the land to agriculture. Remaining
colonies were plagued by rats and pigs. One survey done during the 1960s reported
that less than 5 percent of the petrel nests actually fledged young; the remaining
eggs and chicks had fallen prey to rats, pigs, dogs, and cats. In the 1980s, the Charles
Darwin Research Station and Galápagos National Park Service initiated predator
eradication programs on Santa Cruz and Floreana, and these efforts have succeeded
well, with a reported fledging rate of 80 percent. As is the case with the tortoises and
land iguanas, these efforts, while successful, must not cease. The problem remains
ongoing.

Marine ecosystems are currently the most threatened of Galápagos ecosystems,
even though, in 1986, just over 17 million acres (7 million ha) of ocean were declared a protected area as part of the Galápagos Marine Resources Reserve. There is
little if any evidence that this protection is meaningful. Great pressures are being
brought to bear from international as well as Ecuadorian fishing vessels to harvest
more from the waters around the islands.

It is vital to protect marine ecosystems around the islands. Resources essential to
Galápagos animals are the products of the ocean. It is obvious that animals from sea
lions to marine iguanas to the numerous seabird species all depend upon the sea
to survive. Beyond that, the marine ecosystems themselves comprise complex food
webs. Galápagos coral reefs, various bony fish, sharks, rays, sea turtles, and myriad
invertebrate animals are to various degrees interdependent on each other. Threats
to the oceanic ecosystems quickly become threats to the islands themselves. Ecology does not recognize boundaries between interrelated ecosystems.

One of the most publicized struggles in the Galápagos in recent decades was focused on a marine animal with the scientific name *Isostichopus fuscus,* commonly
called the sea cucumber. Sea cucumbers, which are echinoderms (spiny-skinned animals) related to sea stars and sea urchins, are not normally aesthetically pleasing to
people. They look rather like lumpy, overweight worms. They are deposit feeders,
slurping up organic matter from the ocean mud, a feeding ecology not dissimilar
from what earthworms accomplish as they methodically process food from soil and,
at the same time, aerate and turn over the soil. When threatened, sea cucumbers are

in the habit of forcefully everting their viscera via the anus, a strategy that apparently works in discouraging would-be predators. But, while that defense might work against natural predators, it fails against humans. In places such as Taiwan and Japan, people use the dried flesh of sea cucumbers (called *trepang* or *beche-de-mer*) to thicken and flavor various soups. And that is a problem for the Galápagos.

Under the rules of the Galápagos Marine Resources Reserve, commercial fishing is prohibited from the waters around the islands, but what is called *traditional-style fishing* is permitted in certain zones around the islands. In some zones, all fishing of any kind is, at least in theory, prohibited. But with the economic pressures of Asian markets, there has been a vigorous attempt mostly by the local fishing sector to extract sea cucumbers from Galápagos waters wherever it could find them. One report suggested an astonishing 130,000 to 150,000 sea cucumbers taken per day! Such an exploitation rate will cause the extinction of this animal within a matter of years. Exactly such a result has occurred where intensive fishing of sea cucumbers has been done elsewhere, in places such as Micronesia.

The sea cucumber fishery in the Galápagos is most definitely not part of "traditional fishing." Ecuadorians do not consume trepang. It was not until 1988 that sea cucumbers from Galápagos waters were actually harvested, and this fishing accelerated with such speed that concerns for the health of the marine ecosystems (to say nothing of the sea cucumber itself) resulted in a total ban on sea cucumber harvesting in 1992.

End of story? Only the beginning. New regulations were enacted in 1994 allowing for a minimal harvest of sea cucumbers, a mere 550,000 to be taken over a three-month period. But demand for the animals in Asia pushed prices up, which was more than sufficient to motivate the Ecuadorian fishing industry to ignore the quota and take as many sea cucumbers as it could get. So, as of December 15, 1994, another total ban on sea cucumber harvesting was instituted. It had little effect. Locally the ban was ignored, and there were even reports of camping on such pristine islands as Fernandina and parts of western Isabela. Tourists observed fishermen on the beaches of supposedly protected areas. Beyond that, social unrest among the resident and migrant fishing sectors resulted in, of all things, a brief takeover of both the Galápagos National Park Service offices and the Charles Darwin Research Station, with open threats to damage the various conservation programs and to actually harm the research station staff and animals. In recent years, some irate fishermen actually took some endangered tortoises "hostage" in an attempt at pure intimidation. Of course the angered parties also threatened to disrupt tourism as well. In 1997 a park ranger was shot, and in an attempt to appease the fishing sector, yet another assault on the sea cucumbers was permitted. When the ban went back into effect, a substitute lobster fishery was offered in place of the right to take sea cucumbers. What this apparently resulted in was a major loss of lobsters and, not surprisingly, the continued illegal harvest of sea cucumbers.

Besides pressures exerted on sea cucumbers, sharks are being killed for their fins, and monofilament nets, long lines, and large seines used to catch sharks in protected areas have threatened a collective assemblage of fish and other marine creatures. These affronts were committed by owners of commercial fishing craft (reported to be mostly Asian), none of whom have any legal rights to fish within Galápagos waters. Such trespass is in total violation of the Galápagos Marine Resources

Reserve laws. Be that as it may, there is evidence of interest in greatly expanding commercial fishing operations around the islands, with proposals to build such things as a deep-water dock and a processing plant. Of course such proposals state that all fishing would occur outside of any protected marine areas, but given the lack of adequate patrol and protection, such assurances seem questionable at best.

But there is reason for some hope. On November 24, 1999, the Galápagos National Park Service and the Ecuadorian Navy detained a fishing boat caught only six nautical miles northwest of Wolf Island, well within the Galápagos Marine Resources Reserve. The boat had a long line out, about 20 miles (32 km) worth, and it required about six hours to haul in the line. The catch included numerous sharks, manta rays, and other fish, some of which were alive and were returned to the sea. The boat, the *Mary Cody*, as well as several smaller boats associated with it, were taken to Puerto Ayora. The boat's hold contained many shark fins and shark bodies, all destined for Asian markets. The boats' crews and owners were accused of violating the Special Law for the Conservation and Sustainable Development of the Galápagos.

The waters of the Galápagos face yet another threat, sport fishing. This activity has been proposed by the tourism industry, which largely sustains the islands' economy. The target fish are dolphinfish *(Coryphaena hippurus)* and tuna *(Euthynnus pelamis* and *Thunnus albacares)*. Given the apparent failure to effectively control the sea cucumber harvest, it is certainly dubious as to whether a sport fishery can be managed such that the health of the target species is maintained. The Charles Darwin Research Station is strongly against instituting a sport fishery in the waters around the islands.

The Galápagos are, of course, part of the country of Ecuador, and as such their fate is inextricably linked with that of the mother country. As of this writing, there is some cause for concern for the long-term future of the islands. Ecuador has experienced severe economic stress: reduced export prices, high foreign debt, and an internal banking crisis. Currently the growth rate of the human population in Ecuador is about 2.3 percent annually, making it the fourth-fastest-growing nation in South American. Only French Guiana, Paraguay, and Bolivia exceed Ecuador.

But much worse is the overall growth of human population on the Galápagos, a whopping 7.8 percent annually. This rate exceeds that of any of the African nations. Of the 7.8 percent, about 1.7 percent is from reproduction on the islands, and 6.1 percent is due to net immigration. At this rate, the time it will take for the human population to double on the Galápagos is a mere seven to twelve years. By 2003 the population will be around twenty thousand, and by 2027 that figure will increase four times. Unless the rate of immigration to the Galápagos is curbed, the wildlife on the islands faces a dubious future.

The size of the human population is one hazard to the islands. The residents' attitudes are another. Many of the newcomers to the islands are described as "floating immigrants," who only spend part of the year on the islands and the remainder in Ecuador. In addition people who have arrived within the past two decades dominate the human population of the islands. There is, most regrettably, no evidence that these immigrants came to the Galápagos with any kind of sophisticated understanding of conservation biology. Rather they view their new habitation as an extension of Ecuador, where they do what they must in order to make a living and give no regard whatsoever to the native flora and fauna. The fact that the Galápagos is a national park counts for little in their day-to-day lives. To make matters worse, local schools do not apparently emphasize the unique quality of the Galápagos but, instead, adopt the same curriculum as would be found in any Ecuadorian school on the mainland. Immigration continues to be encouraged in that the Ecuadorian government substantially subsidizes essential human services such as sewage disposal, electricity, and even air transport to and from the mainland.

And there are other problems. On August 1, 1995, the Ecuadorian National Congress passed an astonishing law at the urging of Eduardo Veliz, the elected deputy for the Galápagos. It put control of the Galápagos National Park administration in the hands of politicians, it altered the definition of national park to permit the sustainable use of the park's natural resources by the local population, and it included provisions for the eventual expansion of economic development. It actually required all tourists to the islands to spend at least one night ashore in one of the hotels, and it had no provision whatsoever for prohibiting the introduction of nonnative animals and plants. Fortunately this law was vetoed by the president of Ecuador and never took effect. But the rejection of the law precipitated another crisis similar to that following the sea cucumber ban. Again, there were open threats against the Charles Darwin Research Station and the tourism industry. Again, nothing much came of it in the short term, but in the long term, events like this one are extremely ominous.

On March 18, 1998, Ecuador passed what is called the Special Law of Galápagos, a law that designated the Galápagos Marine Resources Reserve as a fully protected area. This law has not resulted in the cessation of political pressure by the fishing industry, and it remains unclear as to how effective the law will actually be.

During the week of November 13–17, 2000, there was a series of violent incidents directed against conservation biologists by the fishing sector. What prompted the incidents was the fact that the fishing industry had reached the agreed upon 50-ton quota for capture of lobsters in a mere two months when it was permitted to take four months. The government of Ecuador, to appease the fishing industry, actually agreed to raise the quota by an additional 30 tons, a decision that did not sit well with conservationists. Nonetheless, fishermen who had resisted measures to monitor their catch initiated some violent actions directed against researchers and, to a lesser extent, tourists. Some scientists had their offices vandalized, their data destroyed, and even threats of death directed at them, while slogans such as "viva los pescadores" were spray-painted on the walls of labs and offices. Most of the violence, what has been called "ecological terrorism," seems to have been committed by a relatively small number of militants, and the Ecuadorian government arrested and charged three fishermen with terrorism. The efforts of the government notwithstanding, Ecuador remains under pressure not only from the sea cucumber– and lobster-fishing industry but also from those who wish to open all of the Galápagos to long-line fishing and shark fishing.

Paul Watson, president of the Sea Shepherd Conservation Society, said, "If we can't save the Galápagos, what the hell can we save?" Nonetheless, the future of effective conservation of the Galápagos Islands' terrestrial and marine ecosystems is uncertain.

There is a characteristic of the Galápagos that has engendered comment from virtually all enlightened persons who have visited these unique islands. Melville wrote of it, as did Darwin: The animals of the archipelago are extraordinarily tame. In spite of a history that has included major depredations of tortoises and land iguanas as well as the killing of numerous sea lions, these creatures of the Galápagos, and virtually all others, remain essentially fearless of human beings. But how human beings choose to act toward them will most likely determine their fate—and the fate of most other forms of life on Earth.

The Garden of Eden was portrayed as a peaceable kingdom, a place where the birds and beasts lived in harmony, not disturbing and undisturbed by their fellow human inhabitants. There is really no place on Earth other than on the Galápagos Islands where a human being can walk so freely among an animal community. It is this extraordinary tranquility that makes the Galápagos so special. Evolutionary biology can be studied elsewhere and has been. But only on the Galápagos can you see the tenets of evolutionary biology so clearly demonstrated while at the same time you bear witness to what the principal responsibility of humanity should be in this still new millennium: respect for and stewardship of all the creatures of the planet. It is a lesson as profound and as challenging as any in humanity's history. If the lesson is learned, the Galápagos may indeed deserve recognition as a true Eden.

Selected References

Eckhardt, R. C. 1972. Introduced plants and animals in the Galápagos Islands. *Bioscience* 22:585–90.

Hamann, O. 1975. Vegetational changes in the Galápagos Islands during the period 1966–73. *Biological Conservation* 7:37–59.

Hayashi, A. M. 1999. Attack of the fires ants. *Scientific American* 280:26–28.

Loope, L. L., O. Hamann, and C. P. Stone. 1988. Comparative conservation biology of oceanic archipelagos: Hawaii and the Galápagos. *Bioscience* 38:272–82.

Lubin, Y. D. 1984. Changes in the native fauna of the Galápagos Islands following invasion by the little red fire ant, *Wasmannia auropunctata*. *Biological Journal of the Linnean Society* 21:243–51.

Sitwell, N. 1993. The grub and the Galápagos. *New Scientist,* 11 December, 32–35.

Stone, R. 1995. Fishermen threaten Galápagos. *Science* 267:611–12.

12

The Islands

This chapter provides a brief description of the overall natural history of each of the thirteen major islands, as well as some of the smaller ones. These are the islands usually visited by Galápagos tour groups, and this chapter provides a convenient orientation to each island so you will know what to expect if and when you visit. The islands are considered in order from largest to smallest.

Isabela

Origin of Name. Named for Isabela de Castilla, Queen of Spain.

Original English Name. Albemarle (named for the Duke of Albemarle of England).

Official Name. Isabela.

Commonly Used Name. Isabela.

Area. 1,771 square miles (4,588 km²).

Highest Elevation. 5,599 feet (1,707 m).

Notable Characteristics. Very large island (about 80 miles long, or 129 km) with six prominent shield volcanoes, five of which are considered active, including Vulcán Wolf, the highest volcano on the archipelago; extensive lava fields easily visible; volcanic craters (Alcedo, Azul, Darwin, and Wolf) each supporting a distinct subspecies of giant tortoise, with a total tortoise population of around six thousand, the largest of any of the islands; hikes (strenuous) possible to Vulcán Alcedo to see tortoises; flightless cormorants and Galápagos penguins easily seen on the rocks near Tagus Cove; one of the four Galápagos Islands visited by Darwin.

Prominent Landing Areas and Anchorages. Tagus Cove, Tortuga Bay, Punta Moreno, Cartego Bay, Shipton Cove, Puerto Villamil. Also Black Beach for mangrove finch.

Settlements. Puerto Villamil, Santo Tomás.

More about the Island. For most visitors, the first close-up view of Isabela is Tagus Cove, a historic pirate harbor at Punta Tortuga, on the western side of the island. This was the place where Charles Darwin set foot on Isabela back in 1835. Just to the west of Tagus Cove, across the narrow Bolivar Channel, is Fernandina, the youngest Galápagos island. Isabela is an elongate, roughly J-shaped island oriented primarily north to south for a total distance of about 80 miles (129 km). It has six massive volcanoes and is almost five times larger in area than any other Galápagos island. It contains in excess of half the total land area of all the islands taken together. The Equator passes directly through the cone of Vulcán Wolf, the northernmost of the island's volcanoes and highest point on the island. The island consists of a northern region with four volcanic peaks and a southern region with two peaks. The two regions are connected by the Perry Isthmus, with Elizabeth Bay to the west and Cartago Bay to the east.

Tagus Cove is bordered by steep escarpments of volcanic tuff that support enough vegetation to color the surface light green during the wet season. An abundance of palo santo trees covers the steep slopes leading inward from the coast. Where the tuff is exposed, there is scattered graffiti contributed by numerous visitors, an unattractive tradition that goes back many years and many mariners.

Tagus Cove and Bolivar Channel attract numerous seabirds. Blue-footed boobies are usually diving for fish, Elliot's and Madeiran storm petrels occasionally dart around the channel, magnificent frigatebirds sail effortlessly overhead, and small numbers of brown noddies and brown pelicans fish the cove. It is enjoyable to take a panga ride along the rocky cove to look for two of the most remarkable birds on the Galápagos, the flightless cormorant and the Galápagos penguin. The black rocks, densely populated by Sally Lightfoot crabs, are frequently penetrated by volcanic vent tubes. In some areas, numerous boulder-sized volcanic rocks are imbedded in a matrix of hardened material, looking like huge stones tossed in cement before it had solidified. Rocky overhangs and sheltered grottos, some whitened by bird guano, provide nesting sites for seabirds and Galápagos martins. Nests of blue-footed boobies and brown pelicans can usually be seen. Brown noddies nest on the narrowest of ledges, and the sitting birds can be very difficult to spot, so well do their colors blend with the background of lava rock. Flightless cormorants can be seen diving for fish from the water's surface or standing idly on the rocks at the water's edge, occasionally spreading their undersized, vestigial wings. These large birds are easily approached to a distance at which their bright turquoise-colored eyes are visible. Their short brownish body feathers resemble fur. Flightless cormorants are common along the western side of Isabela, not only around Tagus Cove but also all the way from Punta Albemarle in the extreme north to Punta Moreno in the south. Small groups of Galápagos penguins also frequent the grottos around Tagus Cove. They are often observed standing on rocks side by side with a few marine iguanas. Penguin colonies can be found on several islets in the inner part of Elizabeth Bay, south of Tagus Cove.

There are few pinnipeds to be seen around Tagus Cove, but on the eastern side of Isabela, fur sea lion colonies can be found.

Tagus Cove provides a site for a safe, dry landing on volcanic rocks reinforced with concrete. There is a trail that climbs steeply past a rounded crater lake named after Charles Darwin and on up to an overlook permitting a stunning view of expansive lava flows to the northeast. At the pinnacle of the trail there is a panoramic view of Vulcán Darwin directly to the east. The trail itself is composed of fine volcanic rock, very porous, light in weight, and rough textured. Lava tubes line Isabela, and both it and neighboring Fernandina show extensive evidence of recent volcanic activity, with large areas of lava flow. For instance, the entire eastern side of Vulcán Alcedo consists of a vast pumice field.

Of the six major volcanic cones on the island, only one, Vulcán Ecuador, located on the northwestern part of the island, is considered inactive. There have been regular eruptions on Wolf, Sierra Negra, and Cerro Azul throughout the twentieth century. Vulcán Azul erupted as recently as 1998. Vulcán Alcedo, to the south of Tagus Cove, has periodically active geysers and fumaroles. When anchored in Bolivar Channel, it is possible to see Vulcán Wolf to the north, Vulcán Darwin (4,350 feet, or 1,327 m) directly east, and Vulcán Alcedo (3,700 feet, or 1127.8 m) to the south. Sailing south, from Bolivar Channel into Elizabeth Bay, the southern region of the island dominates, with Vulcán Cerro Azul looming at an elevation of 5,540 feet (1,688.6 m) and Vulcán Sierra Negra at 4,890 feet (1,490.5 m).

Commonly occurring along the trail are yellow-flowered *Cordia lutea,* locally called *muyuyo,* a shrubby tree highly characteristic of the arid and transition zones. The leaves are thick and rough, almost like sandpaper, an adaptation to reduce water loss. By far the most common trees are the palo santo, which are abundant from sea level all the way up the trail. These short pale-barked trees are naturally widely spaced, giving somewhat the look of a huge orchard planted along the slopes. Lining the trail are dense clumps of an alien grass that has seeds in sticky capsules called *awns* that will easily attach to socks and shoe laces. Visitors must carefully remove these seeds before leaving the island, so as not to transport them to other islands, especially Fernandina, which, thus far, has been spared this invader.

Land birds are conspicuous. Galápagos martins skim the air for insects. Galápagos hawks often soar in the updrafts created by the volcanic mountains. Many yellow warblers can be found along the trail, and Galápagos flycatchers are common. Darwin's finches abound, particularly small and medium ground finches. Their bulky nests can frequently be seen in the palo santo trees along the trail.

Some Galápagos tours offer the opportunity to ascend Vulcán Alcedo to see the tortoise population that lives within the wide cone. The volcano can be approached by landing near Shipton Cove, a wet landing on the eastern side of the island. Fur sea lions can usually be seen in this area. The hike up to the crater is strenuous and is normally accomplished by camping overnight. The trail to the caldera passes through several vegetation zones and includes good views of lava flows, a fumarole, and the caldera floor itself. In addition to the tortoises, there are land iguanas *(Conolophus subcristatus)* on Alcedo. Land iguanas can also be found on Darwin and Wolf.

For the serious bird-watcher, a visit to Black Beach, located north of Tagus Cove, at Punta Tortuga, is strongly recommended. For it is only here that you can search, with some confidence of success, for the rarest of Darwin's finches, the mangrove finch. Black Beach is so named for its substrate of black volcanic sand, which is generously mixed with shell fragments. Punta Tortuga is also well named—there were numerous sea turtle tracks on the sand when I visited in May of 1997. A short distance from Black Beach is a dense though small (in area) forest of mangroves, mostly red and black. When I was present, there was much bird song in the mangroves, mostly from small and medium ground finches, Galápagos mockingbirds, and yellow warblers, the last extremely abundant. Along the edge of the mangroves was a mangrove finch, a breeding male with a black bill, gathering nest material. Eventually we located a nest of mangrove finches high in a mangrove. We saw and heard the pair going to and from the nest and watched the birds as they foraged, nuthatchlike, occasionally upside-down, mostly on trunks and thick branches. This species looks and acts very much like the woodpecker finch, its nearest relative among the Geospizinae.

On the southern part of Isabela from west to east, the beaches are gentle, composed mostly of coarse sand ground from shells. This is an area of highly saline soil, with scattered salt lagoons that provide habitat for flamingos and white-cheeked pintails, as well as herons, egrets, and various shorebirds. There are extensive mangrove forests interspersed along the southern coast, particularly near the settlement of Puerto Villamil.

There are two modest settlements on the island, Puerto Villamil, a coastal fishing village, and Santo Tomás, an inland farming community that once was a center for tortoise hunting. Because of its history of human habitation, the island has scattered populations of feral cats, dogs, cattle, and donkeys.

Santa Cruz

Origin of Name. Name means holy cross.

Original English Name. Indefatigable (named for the HMS *Indefatigable*).

Official Name. Santa Cruz.

Commonly Used Names. Santa Cruz.

Area. 380 square miles (986 km^2).

Highest Elevation. 2835 feet (864 m).

Notable Characteristics. One of the two most inhabited islands on the Galápagos with a large human settlement at Puerto Ayora on Academy Bay; location of the Charles Darwin Research Station; diversified plant communities with typical Galápagos vegetation zones clearly evident with changing elevation; much farming inland, especially around Puerto Ayora; easily traversed road leading from Puerto Ayora to higher elevations, passing through various ecological life zones; wild domed shell tortoises at a reserve on the southwest side of the island and occasionally outside the reserve; woodpecker finch found in suitable habitat, as is the Galápagos crake; various subspecies of giant tortoises (in captivity) as well as vegetarian finches and large-billed ground finches found at the Charles Darwin Research Station; land iguanas found around Conway Bay on the northeastern side of the island; excellent mangrove forest, sea turtles, and golden and cownose rays found at Black Turtle Lagoon, on the north side of the island.

Prominent Landing Areas and Anchorages. Academy Bay, Conway Bay.

Settlements. Puerto Ayora, Bellavista, Santa Rosa.

More about the Island. Santa Cruz Island is located due east of Isabela to the southeast of San Salvador. The island is rounded and has about twenty small- to moderate-sized volcanic cones scattered mostly in the central region, somewhat like a belt from west to east. There are no large periodically active volcanoes as are found on Isabela. The island is considered to be composed of two geologically distinct regions, the older of which is a thin band along the northeastern coast composed of sandstone and old submarine lava flows and tuffs. Fossils can be found within the sandstone. The vast majority of the island is younger, composed of basaltic lava. The island presents an excellent opportunity for the naturalist to see the eight species of Darwin's finches that occur on the island and the different vegetation zones that characterize the Galápagos.

Academy Bay at Puerto Ayora is by far the most frequented anchorage of the islands and supports a permanent human population of over three thousand that is growing annually. All tour vessels as well as many private craft from all over the world stop here. Puerto Ayora closely resembles a typical Ecuadorian small town, such as would be seen on the mainland. The narrow paved streets are thoroughfares for buses, bicycles, a few automobiles, and other conveyances. The town features a post office, a local radio station, a bank, a bakery, a pharmacy, a hospital, an airline office, lots of small shops where souvenirs and postcards are sold, and several restaurants, some with colorful local names such as Woodpecker Finch and Prosperity. There are also about a dozen small hotels and pensions, including the Hotel Darwin, in which, incidentally, its namesake never stayed. One mile to the east of town is the Charles Darwin Research Station, also on the coast. It is an easy and enjoyable walk from town to the research station.

Ships enter Academy Bay from the southeast, passing tiny Coamano (Jensen) Island, where there are many marine iguanas in residence. To the west, Punta Estrada juts out and forms a steep, protective wall for the harbor. Punta Estrada is dominated by lowland, arid vegetation, with an abundance of the cacti genera *Jasminocereus* and *Opuntia*. Punta Estrada's cliff face is thoroughly painted with graffiti and streaked with bird excrement. Mangroves are abundant along the shoreline, populated by yellow-crowned night-herons, lava herons, and yellow warblers. Blue-footed boobies and a few brown pelicans are usually plunge diving in the harbor.

A ride into the highlands of Santa Cruz is one of the best (and easiest) ways to see a transect of the ecological life zones that occur throughout the islands. Because climate changes at different elevations, it becomes decidedly damper as the road climbs into the highlands. It is wise to bring along rain gear as protection against the effects of the omnipresent garúa mist, which predominates in the highlands at certain times of the year. The bus usually climbs fairly slowly on the unpaved roads, giving lots of time to observe.

The primary terrestrial ecosystem along the coast of Santa Cruz is an assemblage of numerous palo santo trees, dense stands of the *Croton* species, and matazarno bushes, as well as many tall cacti, giving a distinctly arid look to the landscape. This barren appearance begins to change quickly once some elevation is gained. Tall cacti and stands of palo santo give way to agriculture, with fields of imported African elephant grass, plantations of bananas and coffee, and scattered papaya and avocado trees. Cattle, many with attendant cattle egrets, graze in pastures. The reddish soil makes this agricultural region of the Galápagos seem very similar to places on mainland Central and South America. The dense roadside vegetation is often populated by small groups of smooth-billed anis, and it is possible to observe short-eared owls, and sometimes barn owls, occasionally flying low over the fields. Houses and ranches, most looking less than prosperous, dot the roadside but are not abundant.

Farther up in the wetter highlands, the look of the landscape again changes as dense forests of shrubby *Scalesia* species appear. There are eight *Scalesia* taxa on Santa Cruz, most locally distributed. For instance, the species most common in the south near Puerto Ayora are *S. affinis, S. pedunculata,* and two subspecies of *S. helleri.* Of these, only *S. pedunculata* grows as a tree; the others are shrubs. Mixed among the *Scalesia* species are scattered hardwood trees such as the gray-barked guayabillo and reddish-barked guayabo. The scalesia zone is generally moist, and the cultivated areas below are mostly lands that have been cleared of *Scalesia* species. Epiphytes, which thrive in the moist climate, are abundant on the branches of shrubs and trees, especially at higher elevations where the cacaotillo *(Miconia robinsoniana)* joins *Scalesia* species and eventually replaces them in dominance. The epiphyte community includes liverworts and mosses, various fern species, a few orchids, and bromeliads, all supported by the cool garúa mist that often shrouds the higher elevations. Parasitic mistletoe commonly grows on the tree branches. Among the *Miconia* are several species of ferns (most commonly bracken fern) as well as the tree fern *(Cyathea weatherbyana).* Sedges are also common in the wetter areas.

Approximately 10 miles (16 km) from Puerto Ayora, past the towns of Bellavista and Santa Rosa, are two deep pit craters, one on either side of the road. Appropriately named the Twins (Los Gemelos), these craters are surrounded by wet habitat typical of the higher-elevation humid zone. Lush, low-lying species of *Scalesia* and *Miconia,* as well as ferns, surround and grow within the craters. The view into the craters is quite wonderful.

The Twins is an excellent area for birding. Galápagos doves, yellow warblers, and Galápagos and vermilion flycatchers are all common. The wet understory, especially among sedge, is a good place to search for the diminutive Galápagos crake, one of the more difficult endemics to see well. This is also an ideal location to see a diversity of the Darwin's finch species. Many small and medium ground finches can be found as well as small and large tree finches, warbler finches, and woodpecker finches. At this elevation, more like a nuthatch than a woodpecker, the woodpecker finch forages by climbing about the thicker branches searching the bark and epiphytes for arthropod food. Because no cacti are found at this elevation, the woodpecker finch has no source of spines with which to probe bark; thus it does not behave in the way that made this species famous as a tool-using bird.

Southeast of Santa Rosa there is a tortoise reserve where wild populations of the Santa Cruz domed shell tortoise live. These animals can sometimes be observed relatively close to the road, but even though the total population numbers well over a thousand, often they are quite widely scattered and difficult to find. It is wise to ask

One of the Twins' craters in the highlands of Santa Cruz

about recent sightings before embarking on a search. The National Park Service, whose offices are in Puerto Ayora, will grant overnight camping permits for those who wish to camp in the tortoise reserve.

The road continues past the Twins to the northern coast of the island and passes through higher-elevation pampa vegetation, but the drive is not nearly as ecologically varied because it soon crosses mostly through low-lying palo santo stands. At Punta Carrion it is possible to see Galápagos fur sea lions. Where the road terminates at the coast, there is a ferry to nearby Baltra.

Black Turtle Cove Lagoon is a series of small inlets and labyrinthine channels among lush mangrove forest on the north side of the island, southwest of Baltra. The scattered volcanic rubble just outside the lagoon is usually tenanted by brown pelicans, noddies, and blue-footed boobies.

Once inside the cove, the water is shallow, warm, and clear throughout the lagoon. The outermost of the mangroves is almost entirely red mangrove, but other species are found in the interior. Some red mangrove seedlings can be seen dangling like pods from the parent trees. The mangrove prop roots are lined with oysters, and many small concentric pufferfish, the species that gathers around boats (locally called "garbage fish" for this habit), are often swimming among the roots. Many yellow warblers are evident as well as a few smooth-billed anis, some small and medium ground finches, and occasionally lava herons. Noddies and brown pelicans are usually diving for fish. This is one of the best places to see schools of golden rays, also called Pacific cownose rays. They typically glide effortlessly past the panga, almost leisurely "flying" through the calm water. It is also possible to see spotted eagle rays and round stingrays, the latter impressively wide. Then there are the sea turtles. This is one of the best places on the islands to obtain splendid views of several large southern green sea turtles. The transparent water reveals these large animals in full detail as they swim by, sometimes with heads out of the water, affording as wonderful a view of them as they have of you.

Located near Puerto Ayora, the Charles Darwin Research Station, which forms the nexus of much of the scientific and conservation work occurring on the Galápagos, was constructed in 1958, in time for the centennial celebration of the publication of *On the Origin of Species*. The station opened in 1962 under the directorship of Roger Perry and was officially inaugurated in 1964, when it began an ambitious program of hand-rearing giant tortoises to restock islands where these animals had suffered depletion. Restocking first took place in 1970 and has continued since then.

There is a dock at the research station so you can make a dry landing. There is also an information center (with a small bust of Darwin), offices for staff, and the Van Straelen Exhibition Hall, as well as tortoise-holding and -rearing pens.

Neatly kept winding gravel paths (often frequented by lava lizards) take the visitor through the grounds of the research station. Dominant vegetation consists of tall species of *Opuntia* and *Jasminocereus* cacti, as well as shrubs typical of the arid, coastal zone.

One tortoise pen is normally open to visitors, where you can enter and make friends with some of the giant domed shell tortoises. Several of these leviathans seem to enjoy having their long wrinkled necks scratched, extending them upward to gain the most from the experience. A neighboring pen houses some of the least common tortoise varieties, the saddlebacks, including Lonesome George, a male saddleback that is the only known tortoise from Pinta Island. The well-kept pens have ample shade, water, and greenery upon which the inhabitants dine.

There is also a tortoise and land iguana propagation house, a large circular room with about ten separate pens, which hosts baby tortoises and land iguanas hatched from eggs collected on various islands, along with displays explaining about tortoise and iguana biology. The juvenile tortoises look much like very miniaturized versions of the adults, even down to their characteristic resting and standing postures.

The outdoor tortoise pens, and especially the greenery offered as food to the tortoises, attract many bird species. The pens are some of the best places to look for the large-billed ground finch among the much more numerous small and medium ground finches. It is also likely that you will see the small tree finch, and it is possible to see the parrotlike vegetarian finch here. In addition to Darwin's finches, there are swarms of yellow warblers, many Galápagos mockingbirds, and the commonly seen Galápagos flycatchers. In the mangroves near the dock and on the walk back to Puerto Ayora, it is possible to find dark-billed cuckoos.

Fernandina

Origin of Name. Named for Fernando de Aragon, King of Spain.

Original English Name. Narborough (named for Admiral Sir John Narborough of England).

Official Name. Fernandina.

Commonly Used Name. Fernandina.

Area. 248 square miles (642 km²).

Highest Elevation. 4,900 feet (1,494 m), though not precise.

Notable Characteristics. The youngest and most western of the Galápagos Islands; Vulcán La Cumbre still active and erupting periodically; outstanding vista of aa lava fields; large colony of marine iguanas at Punta Espinosa; excellent stands of cacti in the genus *Brachycereus;* well-developed stands of mangrove; good chance of seeing flightless cormorants, penguins, sea lions, and Galápagos hawks.

Prominent Landing Areas and Anchorages. Punta Espinosa, Cape Douglas.

Settlements. None.

More about the Island. At first glance, Fernandina, the westernmost and most geologically recent of the major Galápagos Islands, looks almost like how you would imagine a moonscape. It is largely a gray-black island utterly dominated by volcanic rock, with young spatter cones popping up like so many pimples on a skin of older lava. Vulcán La Cumbre, rising steeply to a height of nearly 4,900 feet (1,500 m), dominates the island, its caldera fully 3,000 feet deep (915 m), its rim extending 4 by 3 miles (6.4 by 4.8 m) across. This volcano is highly active, having erupted over a dozen times since the early 1800s. Surrounding the main crater are numerous smaller craters, also volcanically active. There was once a vast lake on the floor of the crater, but lava flows about thirty-five years ago obliterated much of it. About three decades ago the caldera was strongly affected by an intense series of small earthquakes that caused the floor on the southeastern side to drop almost about 1,000 feet (300 m). Vast tracks of lava extend from the rim of the volcano to the border of the beaches.

As Darwin wrote of this island and neighboring Isabela, "covered with immense deluges of black naked lava, which have flowed either over the rims of the great caldrons, like pitch over the rim of a pot in which it has been boiled, or have burst forth from smaller orifices on the flanks; in their descent they have spread over miles of sea-coast." It is tempting to believe Fernandina is 95 percent covered by lava. Both the ribbonlike pahoehoe (which resembles petrified cow intestines) and rubblelike aa lava are present, the latter much more abundant. The beach at Punta Espinosa comprises black lava sand mixed with shell fragments. Lava flows extend from the beach, almost like jetties, forming habitat for marine iguanas.

The most convenient landing spot on the island is Punta Espinosa, on the northeastern side, across the Bolivar Channel from Isabela Island. This is a dry landing. Trails bordered by stakes keep human visitors from stepping on nests of marine iguanas. The abundant lava lizards take no such heed.

Small forests of mostly red mangrove grow around the point, bordering the sea. The small nonpoisonous Galápagos snake (a *Dromicus* species) can sometimes be found among the dense mangrove prop roots. This is one of the places to search for the mangrove finch, the rarest, most difficult to see, of Darwin's finches. The unique vegetation at Punta Espinosa includes the *Brachycereus* species of cacti, clumps of which are scattered across the otherwise bare lava rocks. These short thick-stemmed cacti are covered by dense stiff spines. They seemingly require no soil base, appearing to sprout directly from the lava itself.

California sea lions are scattered around the beach at Punta Espinosa, many lounging on the sand or swimming, seemingly carefree as they roll about in the large tide pools. Fur sea lions are not common at the point but can be seen at Cape Hammond, on the southwestern side of the island, and Cape Douglas, on the northwestern side.

Sally Lightfoot crabs abound at the point, the juveniles very dark, the older crabs greenish with spotting, and the adults fire engine red.

Marine iguanas are the real attraction of Punta Espinosa. Hundreds are normally present, both adults and juveniles. The marine iguanas on Isabela and Fernandina are the largest in body size of anywhere in the archipelago. They congregate in extensive groups, most pointing their rounded snouts toward the sun, often sitting atop one another. These sprawling reptiles engulf whole sections of rock. Most have snouts coated with whitened, dried salt, the result of processed seawater, which accumulates on their faces when they sneeze. Snorkeling off Punta Espinosa is an excellent way to observe the large lizards as they forage underwater, though, be warned, the waters here are very much on the cold side.

Punta Espinosa is ideal habitat for flightless cormorants. Usually several can be found standing on the lava rocks, preening or gular fluttering, a behavior in which they rapidly vibrate their throat skin to aid in lowering their body temperature through evaporation of water.

It is possible to climb the slope of the caldera of Vulcán La Cumbre, provided the climber is in good physical shape and aware of the risks associated with hiking around active volcanoes. Normally the caldera is reached by following an old lava trail that leads from Cape Douglas, on the northwestern side of the island. Vegetation is generally sparser than that found on older islands: Lava rubble dominates the landscape at higher elevations. Vegetation at low elevations is mostly cactus trees of the genus *Opuntia,* but there are ferns and grasses at midelevation and bushes of the genus *Scalesia* as well as tall grasses at the volcano's crest.

Santiago

Origin of Name. Named for the first island Columbus discovered in the Americas, though not this island of the Galápagos!

Original English Name. James (for King James II of England).

Official Name. San Salvador.

Commonly Used Name. Santiago.

Area. 226 square miles (585 km²).

Highest Elevation. 2,974 feet (907 m).

Notable Characteristics. Extensive lava flows especially at Sulivan Bay; occasional eruptions from small volcanoes; black volcanic sand near James Bay; many goats and donkeys, with evident damage to natural vegetation; Galápagos fur seals and California sea lions in abundance; Galápagos doves; Galápagos hawks; large-billed flycatchers; marine iguanas; abundant lava lizards; one of the four Galápagos Islands visited by Darwin and an island where he did extensive collecting.

Prominent Landing Areas and Anchorages. Espumilla Beach at James Bay, Sulivan Bay near Bartolomé Island.

Settlements. None.

More about the Island. Charles Darwin set foot on Santiago, then called James Island, on October 8, 1835. It was one of only four Galápagos Islands that he was to visit, and he spent a week exploring the island. At that time he dined on tortoise meat. The lumbering reptiles were still in abundance on the island. That situation has regrettably changed because goats and donkeys have essentially taken over.

Two anchorages are frequented, James Bay on the western side of the island and Sulivan Bay on the eastern side (near the tiny island of Bartolomé). For seeing a diversity of wildlife, James Bay is recommended, but for the experience of walking on an utterly immense lava flow, Sulivan Bay is not to be missed.

Santiago is dominated by Cerro Cowan, a large volcano that looms on the northwestern side of the island. Other smaller craters are also evident, some of which are active volcanoes. Santiago has vast areas of lava flows, particularly along its southern side as well as on its eastern side at Sulivan Bay. There is a dry landing spot on the eastern side of the big island. Any doubt about the vulcanism of the Galápagos would be erased after a visit to this site. The island, at this spot, is basically nothing but lava (cooled, fortunately). A walk across it is essentially a primer in lava types, with lots of ropelike pahoehoe among the more jumbled-looking aa. There are some doves, and Galápagos hawks are common here. My group saw a pair of Galápagos hawks in flight, probably courting. A few widely scattered, physiologically hardy shrubs grow among the lava rocks, and a few cacti of the genus *Brachycereus* have a roothold within the metallic rock, but there is no other vegetation. Just lava. The black ropy pahoehoe lava makes a hollow metallic sound as you step on it. Walking across the surface sounds like walking over a graveyard of melted automobiles.

Though some natural vegetation persists on Santiago, the island has been strongly affected by grazing from goats and donkeys, which still persist in abundance. For example, in the region of James Bay there is an abundance of salt sage, a shrub that goats and donkeys strongly avoid.

The island can also be explored from a wet landing at Puerto Egas, on the southern part of Espumilla Beach at James Bay. Near shore in the bay are usually many blue-footed boobies, often in a feeding frenzy, diving like avian bombs into the fertile bay, as well as flocks of frigatebirds, brown pelicans, and brown noddies. This is also a good area in which to see Audubon's shearwaters coursing over the bay. A large volcanic cone named Pan de Azucar (Sugar Loaf) dominates near the beach, with two smaller cones, one to the north and one to the south. Pan de Azucar shows typical vegetation zonation: Slopes are dominated by scattered palo santo trees and muyuyo *(Cordia lutea)*, and its summit is green with *Scalesia* and associated species.

Espumilla Beach is made up of black volcanic sand. Boulders of lava are strewn about, and the beach is bordered by strata created by successive lava flows, like stacks of black metallic pancakes. It is often possible to see tracks of green sea turtles, which come to the beach to lay eggs. The shoreline is generally rocky with many small sinkholes.

Marine iguanas are plentiful along the rocky areas, and California sea lions are abundant. In addition, this is one of the best areas to observe Galápagos fur seals. These animals often spend the sunny hours reclining in the shade created by a small sinkhole, of which there are many to choose. Juvenile and adult Sally Lightfoot crabs are too numerous to count. As for birds, the lava heron, well camouflaged against the volcanic rocks, is common, and shorebirds such as the wandering tattler, American oystercatcher, semipalmated plover, ruddy turnstone, and whimbrel may be encountered.

Away from the shoreline the vegetation is dominated by palo santo trees, scattered grasses, and mixed shrubs, including salt sage *(Atriplex peruviana)*, lantana *(Lantana peduncularis)*, *C. lutea*, and the thorn tree *(Acacia macracantha)*. Large colorful Galápagos grasshoppers can be found as well as dragonflies and spiders of the genus *Argiope*, the latter noted for both their large body and web size.

Not far from Espumilla, near an area called Buccaneer Cove on the extreme northwestern side of the island, there are saltwater lagoons frequented by flamingos and, occasionally, white-cheeked pintail ducks. Galápagos hawks frequent this area, as do Galápagos doves, mockingbirds, and large-billed flycatchers.

San Cristóbal

Origin of Name. Named for Saint Christopher, the patron saint of sailors.

Original English Name. Chatham, after an English gentleman named William Pitt, First Earl of Chatham.

Official Name. San Cristóbal.

Commonly Used Name. San Cristóbal.

Area. 215 square miles (547 km²).

Highest Elevation. 2,939 feet (896 m).

Notable Characteristics. Easternmost of the islands and composed of two fused volcanoes, neither active; second-most-settled island of the Galápagos Islands with second-largest town, Puerto Baquerizo Moreno, the political capital of the province; the small village of El Progreso, established in 1869 (ten years after *On the Origin of Species* was published), the oldest surviving settlement on the islands; airport servicing commercial flights; highland areas, easily accessible by road, allowing a satisfying look at ecological zonation with elevation; El Junco Lake, the only large freshwater lake on the islands; Cerro Brujo, picturesque eroded tuff cone; Kicker Rock, off the western coast of the island; the endemic Chatham mockingbird easily found; one of the four Galápagos Islands visited by Darwin.

Prominent Landing Areas and Anchorages. Kicker Rock (not a landing point but always included in tour itineraries), Cerro Brujo, Ochoa Beach, Sappho Cove, Punto Pitt, El Junco Lake, Frigatebird Hill.

Settlements. Puerto Baquerizo Moreno, El Progreso.

More about the Island. San Cristóbal is often the first island for Galápagos visitors because it now has one of the two airstrips on the archipelago (Baltra has the other). It also has a thriving town, Puerto Baquerizo Moreno, at the southwestern end of the island, which also is the political capital of the Galápagos province. Adjacent to the town is Wreck Bay, where many an ecotour group gets its first look at the boat it chartered for its Galápagos tour.

San Cristóbal serves well as your first Galápagos island. The highland areas abound in vegetation and are frequently shrouded by the garúa. Coastal beaches, such as Ochoa Beach and Sappho Cove, offer ideal opportunities for first-time looks at many of the unique animals, and San Cristóbal is the only place to see the endemic Chatham mockingbird, which is fairly common and, like other Galápagos birds, easily seen. The island is one of the oldest of the archipelago; its volcanoes are now in their geriatric state. The small volcano to the north, Cerro Pitt, was active as recently as two hundred to three hundred years ago, while the larger and more southern of the two craters, Cerro San Joaquin, spewed its last lava something like 650,000 years ago. This impressive volcanic relict has many cinder cones around it, all quiescent.

Most tours include a circumnavigation of Kicker Rock, often at sunset. The rock is so named for its alleged resemblance to a shoe (though it has an alternate name, Sleeping Lion, also for a fanciful resemblance). It is located just west of Sappho Cove, on the western side of the island. Kicker Rock is an old, eroded vertical tuff cone that pokes up from beneath the sea. It emerges abruptly, appearing almost as a volcanic afterthought. The process of erosion has divided the rock into two separate parts, and small tour boats routinely pass between them. The main rock towers over any boat and slopes gently, hosting a mixed seabird colony amidst the soft green vegetation that has a tenuous hold on the rock. Magnificent frigatebirds, blue-footed and masked boobies, and a few brown noddies nest on Kicker Rock, and Elliot's storm petrels are commonly sighted in the surrounding waters. This is a good place to look for waved albatrosses because these birds pass by Kicker Rock as they fly southwest toward their nesting sites on Española.

A paved road runs from Puerto Baquerizo Moreno to the historic town of El Progreso (where the public school has a bust of Darwin and the school is named after him) and continues along the eastern slope of Cerro San Joaquin to El Junco Lake, the largest freshwater lake on the archipelago. This ride takes you from the coastal arid zone up to the highland humid zone, though on this particular island the abundance of introduced plant species and farming activities makes it a challenge to appreciate the native species. Still, the drive is quite scenic, passing through coffee plantations into a mixed scalesia and miconia zone, where the lake is located. Darwin's finches are easy to observe in many places along the roadside, including such uncommon species as woodpecker and vegetarian finches. More commonly you see a mixture of small and medium ground finches, small tree finches, and warbler finches. The picturesque lake is thought to be the remnant of an extinct caldera, now filled with rainwater and daily bathed by the garúa. Sedge, which borders the lakeside, gives the lake its name, El Junco. The lake is an excellent spot to look for white-cheeked pintails and moorhens. In addition, some of the wetter areas, including the meadows, have Galápagos crakes.

On the northeastern tip of San Cristóbal is Punta Pitt, where there is a landing site and dive site situated at the base of an imposing cliff side. Many seabirds can be found here including the swallow-tailed gull and, most importantly, the red-footed booby, which otherwise can only be easily observed on Genovesa, far to the north.

Floreana

Origin of Name. Name means flowers.

English Name. Charles, after King Charles II of England.

Official Name. Santa Maria, after Saint Mary.

Commonly Used Name. Floreana.

Area. 67 square miles (173 km²).

Highest Elevation. About 2,100 feet (640 m).

Notable Characteristics. Most southern of the main Galápagos islands; one of the only islands with dependable source of freshwater; large scenic lagoon that is usually inhabited by flamingos, white-cheeked pintails, and migrating shorebirds; road that goes from Black Beach to highland agricultural area, the only place to see the medium tree finch; home of the Wittmer family, famous in Galápagos history; site of Post Office Bay, where the famous barrel "post office" is set up and still used; outstanding diving and snorkeling at Devil's Crown; Charles mockingbird still seen on Champion Island, but from a boat because there is no access to the island; Green Beach, with greenish-brown sand, and Flour Beach, made up of ground coral; one of the four Galápagos Islands visited by Darwin.

Prominent Landing Areas and Anchorages. Punta Cormoran, Post Office Bay, Black Beach, Devil's Crown (not a landing area but a snorkeling site), Champion Island (not a landing site but the only place to see the endemic Charles mockingbird).

Settlements. Puerto Velasco Ibarra at Black Beach.

More about the Island. Floreana was one of the most oft-visited of the Galápagos Islands in past centuries. Because freshwater was always available, giant tortoises were abundant, and there was a safe anchorage site, sailors frequented the island and established the famous, but odd, barrel post office at Post Office Bay. The release of domestic animals plus depredations by the ships' crews were likely responsible for the loss of the giant tortoise and Charles mockingbird from the island, though the latter can still be seen on Champion Island, a tiny islet just off the northeast coast of Floreana.

Most visitors to Floreana stop first at Punta Cormoran, on the north end of the island. This involves a wet landing on a narrow beach bordered by black mangroves and various scrubby plants. From there it is a short walk to the flamingo lagoon. On the walk it is possible to see small and medium ground finches, as well as Galápagos flycatchers. By far the most abundant passerine is the yellow warbler. Look for the brightly colored males, with red streaking on their breasts and red caps. At the flamingo lagoon there can be over one hundred greater flamingos, sometimes prancing in group courtship behavior and honking like geese or sometimes feeding, their heads upside-down in the shallow saline water. Also in the lagoon are often shorebirds such as semipalmated plovers, black-necked stilts, and whimbrels. White-cheeked pintails are common there, usually seen feeding along the edges of the lagoon. Insects can be a bit of a nuisance, with numerous tiny mosquitoes and some aggressive wasps that seem all too eager to sting. Some of the large black Galápagos carpenter bees are usually evident, and there can be a huge number of flies. Feeding on the insect hordes are numerous large spiders of the genus *Argiope,* whose webs are easily seen on the shrubby vegetation. Palo santo trees abound, lining the slopes around the lagoon, and the volcanic relief of the island is obvious.

A panga ride from the beach at Punta Cormoran to some outlying rocks near a mangrove swamp is excellent for seeing the Galápagos penguin. Cattle egrets roost in the mangroves.

Devil's Crown, near Punta Cormoran, is an abrupt outcrop of rock heavily used by nesting swallow-tailed gulls and red-billed tropicbirds. The so-called crown is, in reality, the jagged peaks of a submerged volcano whose eroded, uppermost part protrudes above the water. Divers love this area for its abundance of hammerhead sharks and other marine life, but the currents can be quite powerful, and divers are advised to take notice. It is not possible to land at Devil's Crown because it is only a dive site.

Champion Island is a small tuff cone very near Floreana, now the last stronghold for the Charles mockingbird. The island, dominated by a rather steep hill, has abundant cacti, some of it the tall *Jasminocereus* species, much of it the *Opuntia* species. There is no landing site, so bird-watchers must look for the mockingbird from the boat. Small boats are permitted to approach the island closely, and with luck, there is little difficulty in finding the mockingbird as well as the more widely distributed cactus finch. Many California sea lions are usually around this area, and some brown noddies are frequently tucked in among the rock crevices.

Black Beach is a dry landing on the western side of the island, south of Post Office Bay. This is the human-inhabited part of Floreana, where the famous Wittmer family still maintains a residence. This is not a thriving community. There is a school and a largely empty assemblage of buildings that apparently pass for a town. From here you can rent a truck or some other form of conveyance and take the only road to the highlands, a road paved with volcanic rubble. The road climbs steadily, leaving the lowland palo santo zone and entering the wetter, more lush scalesia zone. Low, twisted scalesias and numerous other plant species are abundantly draped with lichens and liverworts, making this forest among the most scenic on the islands. Dark-billed cuckoos, vermilion flycatchers, and smooth-billed anis are commonly seen among the small trees and shrubs. Yellow warblers are abundant, as are small ground finches. Less common but easy to find are medium ground finches and small tree finches. It is here, among the scalesia, at elevations greater than 980 feet (300 m), that the rarest bird on the island can be found, the medium tree finch. It occurs nowhere else on the archipelago. As elevation increases, so does the likelihood of rain. Expect it. The road goes into the agricultural zone, where crops such as corn and cassava are raised.

The highland agricultural areas are on the slopes of Cerro Pajas, an old and likely inactive volcano. The last volcanic eruptions reported from this island were in 1813, and lava flows have been dated as far back as 1.5 million years ago. The volcanic hills are eroded, now shrouded in vegetation and constantly soaked in garúa.

Marchena

Origin of Name. Named for Fray Antonio Marchena.

Original English Name. Bindloe, after Captain John Bindloe.

Official Name. Marchena.

Commonly Used Name. Marchena.

Area. 50 square miles (130 km²).

Highest Elevation. About 1,125 feet (343 m).

Notable Characteristics. Dive site only, no landing sites; one large shield volcano, with evidence of recent activity (1992); major fur sea lion colony; no introduced plants or animals save one, a fire ant *(Wasmania auropunctata),* recently arrived.

Prominent Landing Areas and Anchorages. None.

Settlements. None.

More about the Island. Marchena is situated well to the north of Santiago and 31 miles (50 km) west of Genovesa. Because of its relative remoteness, it is not a visitor site, and few go there. About the only thing of real note concerning this island (and it isn't all that much of a note) is that Lorenz, the rejected lover of the baroness of Floreana, apparently died here. At least this is where his corpse was found.

Española

Origin of Name. Named after the country of Spain.

Original English Name. Hood, after Admiral Viscount Samuel Hood.

Official Name. Española.

Commonly Used Name. Española.

Area. 23 square miles (60 km^2).

Highest Elevation. About 675 feet (206 m).

Notable Characteristics. Oldest and one of the most magnificent of the islands; amazing vista from the cliffs at Punta Suarez; blowhole; bright reddish marine iguanas; nesting colony of waved albatrosses; blue-footed booby nesting colonies; nesting red-billed tropicbirds; endemic mockingbirds; large cactus finches; large lava lizards.

Prominent Landing Areas and Anchorages. Punta Suarez, Gardner Bay.

Settlements. None.

More about the Island. Española is considered to be the oldest of the extant islands of the Galápagos Archipelago (some, much older, have eroded and are beneath the sea, essentially "extinct"), with lava flows dated at 3.4 million years. No active vulcanism is present. The island is generally dry and hot.

The most prominent feature of this island is the dramatic cliffs that define its southern coast. The cliffs are reached by walking from the landing point at Punta Suarez, a wet landing on the northwestern end of the island. Almost immediately, inquisitive Hood mockingbirds usually appear, running rather than flying, picking at toes, backpacks, or whatever presumably interests them. As the name implies, the Hood mockingbird is endemic to this island. A trail winds for about 1.5 miles (2.4 km) through low vegetation, passes through a colony of nesting blue-footed boobies, and eventually comes to the waved albatross colony. Galápagos doves are extremely abundant on Española as are small ground finches. But the most interesting of Darwin's finches on this island is the large cactus finch, which, with its huge bill and large body size, looks a great deal like the large ground finch found elsewhere on the archipelago. Both blue-footed and masked boobies nest on the island, but there is a kind of residential "zoning" that they observe: Blue-footed boobies occupy the flat, open areas, while masked boobies nest among the rocks on the cliffs. While walking among the many pairs of nesting blue-footed boobies, you have an ideal opportunity to observe their curious courtship behavior.

The vista that is afforded from the cliffs, where you look down more than 300 feet (91.4 m) to the crashing sea, is unrivaled on the islands. Below, within the complexity of rock fissures, the energy of the sea rises up in the most literal way, as a powerful blowhole periodically throws a picturesque plume of briny water high in the air, almost geyserlike. All of this striking scenery serves ideally as a backdrop for a marvelous spectacle: Masses of bright red marine iguanas, the most colorful of their kind on the islands, climb sheer cliffs to bask atop the rocks; big-eyed swallow-tailed gulls are at arm's length; dozens of masked and blue-footed boobies are perched and in flight, many vocalizing, a cacophony of shrill whistles and honks; small flocks of red-billed tropicbirds speed by, screeching loudly as they perform incredible aerial twists and turns; Audubon's shearwaters soar low over the ocean below; and large rafts of waved albatrosses rest on the sea.

It just doesn't get any better—that is, until a waved albatross, coming from its nesting colony adjacent to the cliffs, waddles up behind you, swaying its huge head curiously from side to side, and stops, waiting for you to get out of its way. When you do, it walks past you, head swaying rhythmically as it goes, to the very edge of the cliff, where it spreads its wings, looks ever so slightly down, and appears to contemplate flight. Once it is satisfied that it has been cleared for takeoff, the ponderous bird drops, its long wings soon catching the updraft, and it stiffly flies out to sea to go about its business. It is very hard not to envy the big bird that evolution made into such a superb oceanic flier. It lives in a wonderful place.

Gardner Bay, on the northeast part of the island, provides a second landing site, with a pristine beach and ideal snorkeling conditions. Many California sea lions congregate in this area, and it is a dependable place for making the acquaintance of the bold Hood mockingbird. Galápagos hawks are also frequently observed here as well as elsewhere on the small island.

Española once held a thriving population of saddleback tortoises that fed on the tree-sized *Opuntia* species of cacti that populate parts of the arid island. Beginning in the 1960s, aggressive captive breeding was conducted with the few remaining Española saddlebacks, and the program has succeeded to the degree that about seven hundred tortoises raised at the Charles Darwin Research Station were released on Española in 1995. These animals are not normally observed by visitors to the island, but if they thrive and increase in number, Española could eventually become one of the best places to see the giant saddleback tortoises.

Pinta

Origin of Name. Named after one of Columbus's original ships.

Original English Name. Abingdon, after the Earl of Abingdon.

Official Name. Pinta.

Commonly Used Name. Pinta.

Area. 23 square miles (60 km²).

Highest Elevation. About 2,800 feet (850 m).

Blowhole at Española

Notable Characteristics. One of the northernmost of the islands; large colony of Galápagos fur sea lions; unique volcanic rocks (abingdonite) containing a white crystalline mineral called *plagioclase;* home of the tortoise Lonesome George.

Prominent Landing Areas and Anchorages. None—not a visitors' island though some boats anchor at the south end of the island and briefly explore.

Settlements. None.

More about the Island. Pinta's only real claim to fame is Lonesome George, the male saddleback tortoise that quite literally represents the last example of his race—if, indeed, he is actually from Pinta (see Chapter 6). Once abundant on Pinta, these tortoises were decimated by sailors, with the coup de grace supplied by competition with goats. The goats have since been eliminated, but the damage was done and the tortoise population too far gone. It is questionable, to say the least, if the one survivor, Lonesome George, can be cross-bred and his offspring (and theirs) raised in sufficient numbers to ever repopulate the island with tortoises.

The island offers sharp relief, with cliffs rising steeply to heights of 300 feet (91.4 m). The island is not considered volcanically active.

Baltra

Origin of Name. Origin of the name uncertain, but thought to be related to military use.

Original English Name. South Seymour.

Official Name. Baltra.

Commonly Used Name. Baltra.

Area. 10 square miles (27 km^2).

Highest Elevation. About 328 feet (100 m).

Notable Characteristics. Airport; facility to refuel boats and ships.

Prominent Landing Areas and Anchorages. Not an island to visit for natural history. Serves basically as a point of embarkation and disembarkation.

Settlements. None, but a ferry crosses the narrow channel between Santa Cruz and Baltra on a daily basis.

More about the Island. No one goes to Baltra for any reason other than that the jetliner lands and leaves from there (though many flights use the airstrip at San Cristóbal) and that a boat must refuel during the course of an island cruise. The airstrip occupies a good part of this flat, arid island located just to the north of Santa Cruz. The airport offers such amenities as a small restaurant and some tourist shops, all at the airstrip. The airstrip was built by the United States during the Second World War. Occupation by the U.S. military resulted in the virtual elimination of the native land iguanas, but a reintroduction (from animals reared at the Charles Darwin Research Station) was begun in 1991 and continues.

Santa Fé

Origin of Name. Named for a city in Spain.

Original English Name. Barrington.

Official Name. Santa Fé.

Commonly Used Name. Barrington.

Area. 9.3 square miles (24 km²).

Highest Elevation. About 850 feet (259 m).

Notable Characteristics. Land iguana species found nowhere else on the islands; magnificent giant cactus forest *(Opuntia echios);* scenic walk from beach up to cliff, affording panoramic view.

Prominent Landing Areas and Anchorages. One wet landing area on the northeastern side of the island.

Settlements. None.

More about the Island. Santa Fé is a small island located to the southeast of Santa Cruz. It is not visited on all Galápagos tours, though it offers a wonderful combination of scenery and unique species. There is but a single landing site, on the northeastern side of the island, where you make a wet landing on a white sandy beach generously covered by sprawling sea lions most of the time. This island is old, with lavas dating back nearly 3 million years, and it consists of two small calderas located well above the beach on the escarpment. The first thing to notice is the dense forest of the treelike cacti of the genus *Opuntia,* the trunks of which are unusually wide. This statuesque forest is unrivaled elsewhere on the islands. Not surprisingly, the cactus finch is abundant here, and nests are easily found. Small and medium ground finches, Galápagos doves, Galápagos mockingbirds, and large-billed flycatchers are also easy to see in the cactus forest. A brief walk through the cactus forest can also reveal the other unique organism of Santa Fé, the pale land iguana *(Conolophus pallidus).* This species, more pale and more uniformly yellow than the other land iguana species, is endemic to Santa Fé.

To find the pale land iguana, it is usually necessary to walk the trail up the escarpment face to the tableland above, a journey that produces one of the truly spectacular views possible on these islands. Fortunately this iguana population seems secure. Goats once represented a problem, but they were eradicated in 1971, and there are no other introduced animals at present.

In addition to the land iguana, there is also an endemic rice rat species on Santa Fé, but few visitors see it.

Pinzón

Origin of Name. Named for the Brothers Pinzón.

Original English Name. Duncan, after Admiral Viscount Duncan.

Official Name. Pinzón.

Commonly Used Name. Duncan.

Area. 7 square miles (18 km²).

Highest Elevation. About 1,500 feet (458 m).

Notable Characteristics. Not a frequently visited island; active tortoise reintroduction program following eradication of black rats.

Prominent Landing Areas and Anchorages. None, landings are difficult.

Settlements. None.

More about the Island. This infrequently visited island is located directly west of Santa Cruz. The island is arid, mostly shrubs, but does support a population of giant tortoises, which is being augmented by introductions of additional animals raised at the Charles Darwin Research Station.

Genovesa

Origin of Name. Named for Genoa, Italy, Columbus's birthplace.

Original English Name. Tower.

Official Name. Genovesa.

Commonly Used Name. Tower.

Area. 5.4 square miles (14 km²).

Highest Elevation. About 255 feet (76 m).

Notable Characteristics. Outstanding island for observing seabirds; large red-footed booby colony; large storm petrel colony; masked boobies, great frigatebirds, swallow-tailed gulls, and red-billed tropicbirds; excellent place to see the sharp-billed ground finch and large-billed ground finch; Galápagos fur sea lions; excellent snorkeling along cliff face.

Prominent Landing Areas and Anchorages. Darwin Bay beach, Prince Phillip's Steps.

Settlements. None.

More about the Island. Genovesa is one of the few Galápagos Islands that is still commonly referred to by its former English name, Tower. It is located well to the north of the main islands, essentially between Santa Cruz and San Cristóbal. A sail to Tower from Puerto Ayora, almost 90 miles (145 km), requires at least eight hours, and thus many tours do not include this island. That is regrettable because it is one of the most wonderful islands of the archipelago. The opportunities for observing seabirds are almost unrivaled here, with the possible exception of Española (Hood).

Tower is shaped rather like a croissant: It has a crescent shape, with a rounded bay that opens to the sea. The bay, named Darwin Bay (although he never visited this island), is actually the center of the caldera because Tower is a single inactive volcano emerging from the sea. A second caldera, much smaller, is found toward the center of the island. Geologists find Tower of interest because the magma that characterizes it is representative of that found along midocean ridges, where new sea floor emerges. The acronym MORB (mid-ocean-ridge-basalt) is used to identify this lava.

Once anchored in Darwin Bay, the island is explored by a wet landing on a small beach of coral sand, also named after Darwin. Some California sea lions are usually lounging about on the sand, but the real show is the seabirds. The island is entirely within the arid zone. In the low shrubs there are nesting great frigatebirds, the males with fully expanded, brilliant red throat pouches, along with red-footed boobies, mostly brown phase but also some white phase birds. This is one of the best places to see swallow-tailed gulls on nests. Some have eggs, some tiny chicks, some chicks that are half or more grown. Masked boobies are numerous. There are also lava herons and yellow-crowned night-herons. The frigatebirds, which whinny constantly, are typically engaged in numerous acts of kleptoparasitism, stealing twigs from boobies, booby nests, and each other.

After spending some time at the frigatebird/booby colony, it is fun to walk along a row of tide pools bordering mangroves to an uneven trail of sharp lava rock. This short trail leads to an overlook where two small towers are located (not the English namesake of the island). The vegetation here is dense, with the *Opuntia* species of cacti the size of small trees. This is an excellent place to see land birds. Galápagos doves are quite numerous, as are the extremely tame Galápagos mockingbirds. This island also features some Darwin's finch species not easily seen elsewhere. These are the sharp-beaked ground finch, the large ground finch, and the Tower race of large cactus finch, which has a bill shaped much differently from the population on Española. Yellow warblers are not quite as numerous as on many other islands, but they are common.

Marine iguanas on Tower are common but quite small compared with those elsewhere on the archipelago.

Snorkeling is excellent along the escarpment on the western side of the caldera. The rocky walls provide good substrate for many fish: the moorish idol, hieroglyphic hawkfish, guineafowl puffer, yellow-bellied triggerfish, yellow-tailed surgeonfish, large-banded blenny, bump-headed parrotfish, streamer hogfish, blue-chinned parrotfish, gray mickey, and king angelfish. While snorkeling, it is also easy to get close looks at Galápagos fur sea lions in the grottos along the wall.

A panga ride along the caldera wall is an excellent way to look for tropicbird nests tucked within the rock crevices high above. There is a dry landing at Prince Phillip's Steps, where His Majesty himself made the climb up the escarpment when he visited the islands several decades ago. The climb up the steep steps takes you to the top of the caldera, where there is a forest of palo santo trees heavily inhabited by red-footed and masked boobies, with ample numbers of spiders of the genus *Argiope,* each perched in its web. Galápagos mockingbirds are also abundant. When I was there, I thought I saw a dead juvenile booby in a tree, its pathetic lifeless head and neck dangling limply, but Luis, our knowledgeable naturalist guide, "resurrected" it merely by walking up to it. It must have really felt the heat or else been terribly bored.

There is a short trail to a flattened volcanic ridge where the labyrinthine rocks (really old lava tubes) make suitable nest sites for many thousands of storm petrels, mostly wedge-rumped. The vista is amazing, with masses of storm petrels coursing about over the colony in swirling, restless, seemingly endless flight. You can even smell them because they emit a strong, musky odor. Careful searching should reveal a pair of short-eared owls among the storm petrels. The owls grow fat on the little black seabirds.

Rábida

Origin of Name. Named after Convent de la Rábida.

Original English Name. Jervis, after Admiral John Jervis.

Official Name. Rábida.

Commonly Used Name. Jervis.

Area. 1.9 square miles (4.9 km²).

Highest Elevation. About 1,200 feet (367 m).

Notable Characteristics. Beach composed of very reddish sand; small saltwater lagoon, often occupied by flamingos and white-cheeked pintails; trail through tall cactus forest for outstanding view from cliffs on southeastern side of the island; abundant cactus finches and Galápagos doves; good snorkeling from red sand beach.

Prominent Landing Areas and Anchorages. Wet landing on sandy beach.

Settlement. None.

More about the Island. Rábida is a small island directly south of Santiago, and it is the geographic center of the Galápagos Islands. It is best known for the distinctive red sand of its beach. The color is the product of a substance that iron oxides in the sand grains. In other words, the sand is literally rusty. Ecologically, the island is entirely within the arid zone. A wet landing is usually greeted by lounging California sea lions on the beach. Myriads of Galápagos doves fly around because this is one of the islands where they are most abundant. There are lots of Darwin's finches near the beach, mostly small and medium ground finches, as well as an abundance of yellow warblers. There is a small lagoon not far from the beach where flamingos can sometimes be found along with an occasional white-cheeked pintail and shorebirds such as the oystercatcher and semipalmated plover.

A loop trail winds through a dense cactus forest (along with other species) to several scenic overlooks that make excellent photo opportunities. The vegetation is dense and supports numerous spiders of the genus *Argiope* and their webs. Cactus finches are easy to observe throughout the walk. Not surprisingly, lava lizards are quite common.

Snorkeling is pleasant just off the beach. You can swim along the rocky island and look into rock crevices for fish and other forms of marine life.

Seymour Norte

Origin of Name. Named for Lord Hugh Seymour.

Original English Name. North Seymour.

Official Name. Seymour Norte.

Commonly Used Name. North Seymour.

Area. 0.7 square miles (1.9 km^2).

Highest Elevation. Basically flat.

Notable Characteristics. Colony of magnificent frigatebirds with some great frigatebirds; large colony of blue-footed boobies; many nesting swallow-tailed gulls; many large California sea lions.

Prominent Landing Areas and Anchorages. Rocks on southwest side allow a dry landing.

Settlements. None.

More about the Island. North Seymour is one of the few Galápagos Islands to retain its English name, though it has been translated, so to speak, into Spanish. But, far more noteworthy, this small flat island of arid zone shrubs, short-stature cacti of the genus *Opuntia,* and dense palo santo trees hosts a thriving colony of magnificent frigatebirds, as well as some great frigatebirds. Both bird species nest in the same area, in low vegetation, affording excellent opportunities for comparing these similar birds. In addition to the mixed species frigatebird colony there are many blue-footed boobies and swallow-tailed gulls. Land birds are not very abundant, but the dense vegetation affords a good chance to encounter the dark-billed cuckoo. The island also has land iguanas, introduced from nearby Baltra in the 1930s, but there are also rats, so land iguanas are not easily seen here.

Bartolomé

Origin of Name. Named after Lieutenant David Bartholemew.

Original English Name. Bartholemew.

Official Name. Bartolomé.

Commonly Used Name. Bartolomé.

Area. 0.5 square miles (1.2 km²).

Highest Elevation. About 374 feet (114 m).

Notable Characteristics. Superb examples of spatter cones; magnificent view from trail summit; moonlike landscape with unusual plants including lava cacti of the genus *Brachycereus* and *Tiquilia* species, which grow on lava soil; Pinnacle Rock; small Galápagos penguin colony at Pinnacle Rock.

Prominent Landing Areas and Anchorages. Dry landing in a small rocky cove.

Settlements. None.

More about the Island. Bartolomé is not to be missed. Tiny but totally splendid, it is one of the premier jewels in the crown that is the Galápagos. The island is widest on the eastern two-thirds, with the western third being a narrow peninsula. The most conspicuous feature is Pinnacle Rock, the remains of an old tuff cone. It is an immense monolith shaped like a spearhead that looms skyward on the north side, where there is a fine sheltered cove with a beach. The air around Pinnacle Rock is typically filled with soaring frigatebirds and diving boobies, as well as brown noddies and brown pelicans, and the whole place has a rather Mesozoic look to it. Most interesting are the Galápagos penguins, small groups of which fish just off the island, affording snorkelers the opportunity to see these birds in action. The penguins also are fond of sitting on the rocks near the cove.

Bartolomé is as volcanic as volcanic can be. It looks rugged, with textbook examples of tuff cones and spatter cones. The volcanic soil is typically dark but has distinct reddish areas. Visitors land at a small dock on the north side of the island and walk a steep trail to an overlook that is perhaps the most photographed place in the islands. The view from the summit (about 375 feet, or 114 m, above sea level) is of lava tubes, lava flows, and numerous irregularly shaped spatter cones that somewhat resemble craters of the moon. Big chunks of old lava are scattered everywhere. From the overlook, you look due west at the beach and cove below, at Pinnacle Rock, and, across the narrow channel of Sulivan Bay, at Santiago and the immense black lava flow that defines its eastern side. This is a wonderful spot to visit at sunset. The naturalist guides will warn you to remain well within the white stakes that mark the narrow trail because the sparse vegetation found here is extremely susceptible to disturbance. The trail to the summit begins with a gradual incline but soon comes to steep steps that lead to the top. Be prepared: I counted 162 of these steps.

Plants have a predictably tough time here and are widely spaced over the "moonscape." Of particular interest are plants in the genus *Tiquilia,* small but relatively abundant gray-green plants that are virtually prostrate, tightly hugging the volcanic soil, eking out a living in what can only be a most inhospitable garden. Four species of *Tiquilia* occur on the islands, all endemic and all in the arid zone. There are only a few other plants to be found: a euphorb *(Chamaesyce amplexicaulis),* which occurs at lower elevations; scattered clumps of lava cacti (members of the genus *Brachycereus*); some thinly scattered palo santo trees along the slopes; and a few scattered treelike cacti of the genus *Opuntia* near the beach.

Lava lizards and the Galápagos painted locust (a large grasshopperlike insect, colored brilliant red, green, yellow, and black) are both common along the trailside. There are not many land birds on Bartolomé, though there are some doves and small ground finches.

Snorkeling is good off of the north side of Bartolomé, and South Beach, a cove on the south side of the island, is known for blacktip and whitetip sharks, spotted eagle rays, and stingrays.

Wolf

Origin of Name. Named for an Ecuadorian geologist.

Original English Name. Wenman, after Lord Wainman (spelling was apparently changed).

Official Name. Wolf.

Commonly Used Name. Wenman.

Area. 0.5 square miles (1.3 km^2).

Highest Elevation. About 830 feet (253 m).

Notable Characteristics. Located far to the northwest of the main archipelago; not a visitor site; sharp-billed ground finches feed on booby blood; doves have larger body size than on main islands and are thought to be a separate subspecies.

Prominent Landing Areas and Anchorages. None, though used as a site for scuba diving.

Settlements. None.

More about the Island. Wolf is discussed next along with its sister island, Darwin.

Darwin

Origin of Name. Named after—you know.

Original English Name. Culpepper, after Lord Culpepper.

Official Name. Darwin.

Commonly Used Name. Culpepper.

Area. 0.4 square miles (1.1 km²).

Highest Elevation. About 540 feet (165 m).

Notable Characteristics. Most northwestern of the Galápagos Islands; large doves; good deep diving (scuba).

Prominent Landing Areas and Anchorages. None.

Settlements. None.

More about the Island. Wolf and Darwin, also known as Wenman and Culpepper, are sort of distant cousins to the rest of the islands of the archipelago. Located about 100 miles (160.9 km) from the other islands, these two tiny islands are sort of outliers, far to the northwest. They usually appear as insets on maps of the archipelago because they are sufficiently remote to make it difficult to depict them accurately on a distance scale.

Though these islands may be distant cousins to the others, they are themselves siblings. Each island is but the top of a very large volcano that originates on the ocean bottom, about 3,300 feet (1,000 m) below sea level. Each island is also part of an elongate, northwest to southeast, volcanic ridge called the Wolf-Darwin Lineament. The lava from these islands has been dated using potassium–argon dating, and Wolf lava is four hundred thousand years old, while that of Darwin ranges from nine hundred thousand to 1.6 million years old.

Although there are no visitor sites on the island, there have been studies of the birds on Wolf, and for good reason. The sharp-billed ground finch that lives there is the so-called vampire finch, so named for its peculiar habit of pecking into the backs of boobies and ingesting blood from the larger birds. This behavior is clearly an example of evolution in action.

Daphne Major

Origin of Name. Named for the HMS *Daphne*.

Original English Name. Daphne Major.

Official Name. Mosquera.

Commonly Used Name. Daphne Major.

Area. 0.12 square miles (0.32 km²).

Highest Elevation. About 366 feet (120 m).

Notable Characteristics. Large blue-footed booby colony within caldera; large tuff cone; island used extensively for ornithological research.

Prominent Landing Areas and Anchorages. None. Landing is difficult.

Settlements. None.

More about the Island. Daphne Major, located northwest of Santa Cruz, is the research station for Peter and Rosemary Grant (and their students), whose work on Darwin's finches was chronicled in the Pulitzer Prize–winning book, *The Beak of the Finch*. The island is representative of arid zone ecology, and the plants are strongly affected by droughts and abnormally wet years, patterns that were responsible for driving natural selection in the medium ground finch and cactus finch populations.

Seabirds abound around Daphne Major: red-billed tropicbirds and frigatebirds, as well as a large blue-footed booby colony in the crater itself and many nesting masked boobies atop the rim and along the slopes of the tuff cone. The island is also a reliable place to find Galápagos martins, which, because the island is not normally open to visitors, can often be seen from boats passing closely.

Daphne Minor is a tiny islet nearby and is not open to visitors. No one goes there.

Sombrero Chino

Origin of Name. Name means "Chinese hat," for the island's shape.

Original English Name. None.

Official Name. Sombrero Chino.

Commonly Used Name. Sombrero Chino.

Area. 0.08 square miles (0.22 km²).

Highest Elevation. About 160 feet (52 m).

Notable Characteristics. Good examples of pahoehoe lava and scoria; rugged landscape; rocky seascape good for finding various shorebirds, including the wandering tattler and American oystercatcher; many marine iguanas; California sea lions.

Prominent Landing Areas and Anchorages. Wet landings at several locations.

Settlements. None.

More about the Island. Sombrero Chino, which takes its name from its symmetrical young volcanic cone (that looks like a traditional Chinese hat), is a geologically youthful tiny island southeast of Santiago. It's a good way to become acquainted with Galápagos geology because it offers quite a range of volcanic lessons. After a wet landing there is a brief walk past volcanic rubble, pahoehoe lava, scoria, and lava tubes, and there is a good view of the Santiago lava field across the narrow channel. The beach consists of mixed coral and volcanic rubble. The island itself looks like a large spatter cone. Along the rocky water's edge are many Sally Lightfoot crabs, marine iguanas, and numerous relaxing sea lion pups. Lava lizards commonly stalk crabs here.

Plaza Sur

Origin of Name. Named for a former president of Ecuador.

Original English Name. South Plaza.

Official Name. Plaza Sur.

Commonly Used Name. Plaza Sur.

Area. 0.05 square miles (0.13 km²).

Highest Elevation. About 75 feet (25 m).

Notable Characteristics. Abundant land iguanas; large California sea lion colony; a tree-sized species of *Opunita;* dense *Sesuvium* species; cliff face abounding in seabirds.

Prominent Landing Areas and Anchorages. Small dock on north side.

Settlements. None.

More about the Island. Plaza Sur is one of the smallest of the islands regularly visited by ecotour boats. It is unique. Located on the eastern side of Santa Cruz, Plaza Sur is the real Jurassic Park of the Galápagos, populated as it is by an abundance of large land iguanas *(Conolophus subcristatus)*. The sprawling, robust lizards are frequently encountered among the densely growing plants of the genus *Sesuvium*, along the trails, among the groundcover, and under the treelike cacti of the genus *Opuntia*. I watched one working its ample mouth purposefully into a fruit of the *Opuntia* species. The tree-sized *Opuntia* species, along with a dense carpet of endemic *S. edmonstonei*, are the notable botanical characteristics of this little, colorful, wonderful island. The land iguanas build their nesting burrows among the plants of the genus *Sesuvium*. The marine iguanas here are quite small in comparison with their terrestrial kin and very black. Lava lizards, ever present, make up the remainder of the reptilian community. Cactus finches abound, as well as small and medium ground finches.

The island is reached by a dry landing on a small dock where there is typically a large gathering of relatively inanimate California sea lions, most of which seem not the least bit interested in moving. You must step gingerly around the creatures, though the naturalist guides do an admirable job of persuading the prone pinnipeds to amble out of the way. Upwards of one thousand sea lions call Plaza Sur home.

A short loop trail crosses the narrow island to a high cliff that looks down on the ocean. This is a great place to see swallow-tailed gulls, often with chicks, as well as red-billed tropicbirds. Sea lions, including some notably large bulls, typically recline at the top of the cliff. Large schools of yellow-tailed mullets can often be seen in the ocean.

Selected References

Boyce, B. 1994. *A Traveler's Guide to the Galápagos Islands*. 2d ed. San Juan Bautista, Calif.: Galápagos Travel.

Constant, P. 1995. *The Galápagos Islands*. Lincolnwood, Ill.: Passport Books.

Jackson, M. H. 1993. *Galápagos: A Natural History*. Calgary: University of Calgary Press.

Index

Page numbers in *italics* denote figures.

Watkins, Patrick, 9–10
Watson, Paul, 179
waved albatross. *See* albatross, waved
Wedgewood, Josiah, 37
Wehrborn, Eloise Bosquet de Wagner, 11
Wenman Island. *See* Wolf Island
whales, 166–168
whaling, 8
white-cheeked pintail. *See* pintail, white-cheeked
white mangrove *(Laguncularia racemosa)*, 31
Wilson, E. O., 56, 127
Wittmer, Margret, 10–12
Wolf Island, 124, 207–208
woodpecker finch *(Cactospiza pallida)*, 47, 136, 137, 138, 140, 142–143, 186, 194

Xylocopa darwini. See carpenter bee

yellow-crowned night heron *(Nyctanassa violacea pauper)*, 158, 159, 185, 203
yellow-tailed mullet *(Mugil cephalus rammelsbergi)*, 166, 211
yellow warbler *(Dendroica petechia aureolla)*, 125, 126, 132, 183, 185, 186, 187, 188, 195, 196, 205

Zalophus californianus wollebacki. See sea lion, Galápagos
Zanclus cornutus. See moorish idol
Zanthoxylum fagara. See cat's claw
Zenaida galapagoensis. See dove, Galápagos
zooplankton, 117, 154, 163, 166, 167